TABLE OF CONTENTS

EVOLUTIONARY THEORY: ANTHROPOLOGY / SCIENCE

INTRODUCTION TO ANTHROPOLOGY

What is Anthropology?

```
┌──────────────────────────────────────┐
│              Key Issues                │
├──────────────────────────────────────┤
│ • What is anthropology?                │
│ • What are the differences between     │
│   the     various     subfields    of  │
│   anthropology?                        │
│ • How is anthropology different from   │
│   other social sciences?               │
└──────────────────────────────────────┘
```

Subfields of Anthropology

Cultural Anthropology

Archaeology

Linguistic Anthropology

Physical Anthropology

Anthropology and other Sciences

SCIENCE AND SCIENTIFIC METHODOLOGY

Steps in the Scientific Process

```
┌──────────────────────────────────────┐
│              Key Issues                │
├──────────────────────────────────────┤
│ • How is scientific data obtained?     │
│ • What is anecdotal evidence and why   │
│   is it problematic?                   │
│ • What is a "scientific fact"?         │
└──────────────────────────────────────┘
```

Anecdotal Evidence

What is a Scientific Theory?

EVOLUTIONARY THEORY: BACKGROUND

Why Study Evolution?

Key Issues

- Why is studying evolution important?

- What barriers hindered the introduction of evolutionary theory?

- How did Erasmus Darwin, Lyell, Malthus, and Lamarck contribute to the acceptance of evolutionary theory?

Barriers

God designs organisms

Species are fixed

The Earth was created recently

Contributions

Evolutionary thought

The Earth is old

Evolutionary mechanism

Population dynamics

Charles Darwin

Key Issues

- What were the significant events in the life of Charles Darwin?

- Why did Darwin wait so long to publish *Origin of Species*?

EVOLUTIONARY THEORY: NATURAL SELECTION

Overview

Darwin's postulates

Key Issues

- What are Darwin's postulates?

- How does evolutionary change occur when the conditions outlined in the postulates are present?

- How is Darwin's proposal for an evolutionary mechanism different from that of Lamarck?

- Why is the phrase "survival of the fittest" an inappropriate way to describe natural selection?

1. The environment limits population growth.

2. Individuals have traits that lead to variation in survival and reproduction.

3. Individuals with favorable traits will be more likely to pass on those traits to offspring.

4. Over time, those traits will become more numerous.

Darwinism vs. Lamarckism

Survival of the fittest

Evidence of Natural Selection

Finches on the Galapagos studied by Peter and Rosemary Grant

Key Issues

- What are the conditions that led to evolutionary change on the Galapagos islands and with the peppered moths?

The Peppered Moth

Details of Natural Selection

Adaptation

The direction of selection

Key Issues

- What is an adaptation?

- What is the difference between directional and stabilizing selection?

- Why were large beaks of the finches not favored after the drought conditions went away?

EVOLUTIONARY THEORY: NATURAL SELECTION

Details of Natural Selection

Rate of evolutionary change

Importance of reproduction

Constraints due to physics and chemistry

Constraints due to changed environments

Constraints since natural selection cannot look ahead

No ultimate goal

No consciousness or learning

Key Issues

- Why does natural selection occur through many small changes?
- Why is reproduction important?
- What are the various constraints on adaptation?
- Why does natural selection have no ultimate goal?
- How is natural selection unconscious?

EVOLUTIONARY THEORY: NATURAL SELECTION

Adaptation Exercise

Discuss each of the following examples from an evolutionary perspective:

> Example: The tortoise subspecies on the Galapagos Islands vary in the shape of their carapaces (shells). Some have "saddleback" shapes, with high openings above their necks. Given that this could increase vulnerability to predators, how and why did this trait evolve?
>
> Answer: Perhaps in the past, food was more abundant on some islands. When tortoises went to other islands with less food, their low-lying shell openings made it difficult to reach enough food. If there was variation in shell openings, those individuals with higher openings could get more food, meaning that they had a greater chance of survival. This led to an increased chance of reproduction since they were living longer. Their offspring would inherit the saddleback trait and would be more numerous in future generations. If food remained limited, those with higher and higher shell openings would be more likely to survive and reproduce. Over many generations, the saddleback carapaces would become the typical trait for that subspecies.

* Many bird species (most notably cuckoos) lay their eggs in the nests of other bird species. A study of rufous bush robins investigated the effects of the presence of an extra egg in their nests. The research revealed that the robins would often eject the eggs from their nests. Why would they engage in such behavior which involves, in effect, killing a baby bird?

* The hammer orchid from Australia grows a red structure attached to the main part of the plant that resembles a female thynnid wasp, who are flightless. This attracts male wasps to the orchid who try to mate with the plant. Why would the hammer orchid evolve such a structure?

* Experimental research has shown that male sticklebacks (a form of fish) prefer larger females as mates. Assume that choosing between different females has costs which may take the form of increased search time or some other energetic requirement. Why would sticklebacks take the time to distinguish between different mates.

* A certain species of termite from French Guiana (*Neocapritermes taracua*) accumulate toxins on their glands during their lifetimes. When threatened, older termites will blow themselves up, thereby releasing their toxins. Why are these individuals, in essence, committing suicide?

* Blue petrels, a type of seabird, are sometimes hunted and eaten by another type of bird, the brown skua. Skua often locate their prey by honing in on calls made by the petrels during the night. Male petrels are particularly susceptible. One study revealed that about 35,000 birds lost their lives to predators on an island that was one square mile in area. Given the risks associated with making calls during the night, why would male petrels engage in this behavior?

EVOLUTIONARY THEORY: MENDELIAN GENETICS

DARWIN'S DIFFICULTIES

Darwin's knowledge

Problems brought about by blending

> ### Key Issues
> - What is "blended inheritance"?
> - Why does blending pose problems for natural selection?

MENDELIAN GENETICS

Background

Gregor Mendel's contribution

Inheritance is particulate

Terminology

Phenotype / True-breeding lines

Dominant / Recessive

Genotype / Allele

Punnett Square

Homozygous / Heterozygous

Genotype ratio / Phenotype ratio

> ### Key Issues
> - What were the major events in the discovery of patterns of inheritance?
> - What is particulate inheritance?
> - How does particulate inheritance solve the problems for natural selection posed by blending inheritance?
> - What are the various terms used in Mendelian genetics?

EVOLUTIONARY THEORY: MENDELIAN GENETICS

Mendelian Genetics Exercises

1. Mendel kept track of several traits of the pea plants he studied. One such trait was seed texture of which there are two alleles (smooth and wrinkled). First, Mendel created "true–breeding lines", plants which always produced the same kind of peas (smooth or wrinkled). He then crossed the smooth and wrinkled lines to create the first generation. He found that all of the seeds were smooth.

   (a) Which trait is dominant?

   (b) He then crossed the members of the first generation to get a second generation (Sw x Sw). Calculate the phenotype and genotype ratios of this generation.

2. Once Mendel created his second generation (from above) he then "back-crossed" members of the earlier generation with the true breeding lines.

   (a) He first conducted the cross Sw x SS. What would be the resulting phenotype and genotype ratios?

   (b) Next he crossed Sw x ww. What would be the resulting phenotype and genotype ratios?

Mendelian Genetics Exercises (continued)

3. Tongue rolling is a trait thought to be controlled by a single gene. If you can roll your tongue upward, then you have at least one dominant allele. I can roll my tongue, but my wife, Carol, cannot. I have two children, Sarah who can roll her tongue and Claire who cannot. What are the genotypes of the four members of my immediate family?

4. How you fold your hands together is thought to be controlled by a single gene. If you fold your hands together and your left thumb ends up on top, then you are expressing the dominant trait. Suppose that a homozygous recessive individual and a heterozygous individual had children. What would be the genotypes and phenotypes of the children they would produce? If they had eight children, how many would you expect to be left-thumb folders?

5. Whether or not you have attached earlobes is thought to be controlled by a single gene. If you have attached earlobes, then you have the recessive trait. Suppose you have attached earlobes, but both of your parents have unattached earlobes. What would be the genotypes of your mother and father? [Hint: First figure out what your genotype is and then attempt to determine what the genotypes for your parents can be. Remember, when your parents mate, they must be able to produce a child with your genotype.]

EVOLUTIONARY THEORY: MOLECULAR GENETICS

The Cell

Eukaryotic cells

Prokaryotic cells

Key Issues

- What are the major differences between prokaryotic and eukaryotic cells?

- What are the major organelles of cells and what are their functions?

Chromosomes

Overview

Structure

Number

Key Issues

- What are genes?

- What is the structure of DNA?

- Why are the bases of DNA considered complementary?

- How many chromosomes do humans have?

Mitosis and Meiosis

Mitosis

Meiosis

Asexual vs. sexual reproduction

Key Issues

- What is mitosis?

- What is meiosis?

- What is the difference between asexual and sexual reproduction?

Protein Synthesis

Overview

Key Issues

- What is protein synthesis?

- Where does protein synthesis take place?

EVOLUTIONARY THEORY: THE MODERN SYNTHESIS

OVERVIEW

Background

Key Issues

- What is the modern synthesis?

- Why did people initially think that the proposals by Darwin and Mendel were incompatible?

SOLUTIONS

Maintaining Variation

Independent assortment

Recombination

Crossing over

Mutation

Key Issues

- What are the sources of variation in populations?

- How can continuous variation in traits occur?

- What is hidden variation?

- How do populations move beyond their original bounds?

Continuous Variation

Single and multiple genes

The environment

Beyond Original Bounds

Hidden variation

EVOLUTIONARY THEORY: THE MODERN SYNTHESIS

POPULATION GENETICS

Overview

When has evolution occurred?

Calculating gene frequencies

Natural selection and evolution

Problems

Gene Frequencies - In human populations in Africa, two common alleles affect the structure of hemoglobin, the protein that carries oxygen on red blood cells. What is regarded as the normal allele (the allele common in European populations) is usually labeled A and the sickle-cell anemia allele is designated S. There are three hemoglobin genotypes. In an African population of 10,000 adults there are 7,000 AA individuals, 3,000 AS individuals, and 0 SS individuals. What are the frequencies of the A and S alleles?

Sickle-Cell Anemia - The three common genotypes associated with sickle-cell anemia have very different phenotypes: SS individuals suffer from severe anemia, AS individuals have a relatively mild form of anemia but are resistant to malaria, and AA individuals suffer from malaria. The frequencies of the A and S alleles are 0.8 and 0.2, and the survival rates for the AA, AS, and SS genotypes are 75%, 100%, and 0%. Will evolution take place from one generation to the next in a population of 1,000 individuals?

Tay-Sachs Disease - Tay-Sachs disease is a lethal genetic disorder which results from the buildup of a molecule in the brain of young children which destroy the brain cells and lead to death by the age of four. The normal allele is labeled N and has a frequency of 0.7, while the Tay-Sachs all, T, has a frequency of 0.3. Assume that survival rates for a population of 2,500 individuals are 100%, 50%, and 0% for the NN, NT, and TT genotypes. Will evolution take place from one generation to the next?

EVOLUTIONARY THEORY: MIDTERM 1 OVERVIEW

GENERAL INFORMATION

The midterm is worth 150 points and covers information presented in the lectures. You must bring a pencil and a Scantron Test form. A bluebook will not be required. There will be 25 objective questions (5 points each) and 2 longer problems (25 points combined). In addition, you must complete the 10-point online portion of the midterm by the date specified on Canvas. You will have the entire class time to complete the exam. There should be very little time pressure. Cheating will not be tolerated.

HELPFUL HINTS

You will be asked to describe natural selection as you did with the in-class exercise.

You will be asked a question on Mendelian genetics.

Sample Question: Darwin waited decades to publish his theory of evolution because he was not convinced he was right (T/F).

Sample Question: Chromosome bases are complementary which means that
   a.   they are polite to one another.
   b.   one can always replace another.
   c.   adenine always binds with guanine.
   d.   thymine always binds with cytosine.
   e.   if you know what base fits on one side of a DNA strand, then you know what base fits on the other side.

PRIMATOLOGY: SPECIES

What is a Species?

The Biological Species Concept

Key Issues

- What is a species according to the Biological Species Concept?

The Origin of Species

Overview

Key Issues

- How are new species formed?
- What processes cause groups to merge back together or diverge?

Merge or diverge?

Diversification

Niche

Key Issues

- What is a niche?
- What is adaptive radiation?

Adaptive radiation

PRIMATOLOGY: SPECIES

RELATIONSHIPS

Overview

Phylogeny

```
┌─────────────────────────────────────┐
│              Key Issues              │
│ ································     │
│ • What is a phylogeny?               │
│ • What are derived traits?           │
│ • What are some derived traits that  │
│   distinguish mammals from other     │
│   groups?                            │
│ • What are convergent traits?        │
└─────────────────────────────────────┘
```

Derived traits

Convergent traits

Phylogeny Example

```
┌─────────────────────────────────────┐
│              Key Issues              │
│ ································     │
│ • What    is    the    evolutionary  │
│   relationship    between    humans, │
│   chimpanzees, and gorillas?         │
└─────────────────────────────────────┘
```

TAXONOMY

Cladistic Taxonomy

```
┌─────────────────────────────────────┐
│              Key Issues              │
│ ································     │
│ • What is taxonomy?                  │
│ • What are the differences between   │
│   cladistic taxonomy and evolutionary│
│   taxonomy?                          │
└─────────────────────────────────────┘
```

Evolutionary Taxonomy

General Phylogenies Exercises

Phylogenies can be reconstructed using genetic data. Instead of basing the phylogenies on shared derived characteristics, overall genetic similarity is used. Below are data for genetic distances among humans and our closest living relatives. Smaller numbers indicate closer relations. Reconstruct the phylogeny of the species.

	Chimpanzee	Gibbon	Gorilla	Human	Bonobo
Gibbon	4.76	–			
Gorilla	2.37	4.75	–		
Human	1.63	4.78	2.27	–	
Bonobo	0.69	5.00	2.37	1.64	–
Orangutan	3.58	4.74	3.55	3.60	3.56

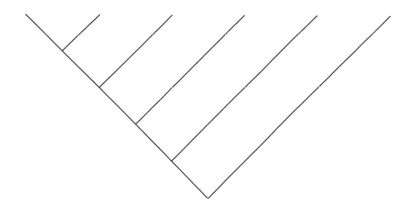

On the following page is a reconstructed phylogeny of selected groups from the animal kingdom. For each of the groups below, indicate which two animals or groups of animals (listed in alphabetical order) are the most closely related. You will need to locate the animals or groups on the phylogeny and identify which share the closest ancestor.

- Chimpanzee (Apes), Human (Apes), Squid
- Baboon (Old World Monkey), Chimpanzee, Dog
- Baboon, Dog, Human
- Eagle, Lobster, Python
- Human, Lobster, Python
- Cow, Gorilla (Apes), Pig
- Human, Kangaroo, Squirrel
- Frog, Human, Oyster
- Human, Lobster, Oyster
- Baboon, Gorilla, Human

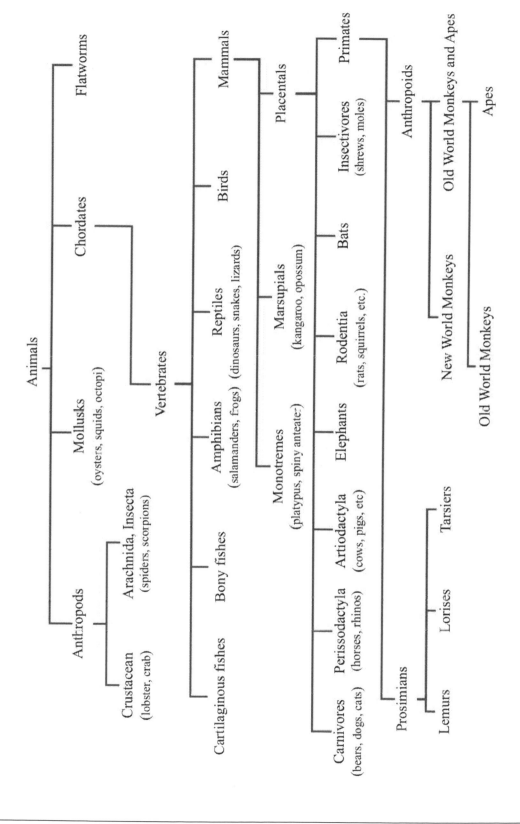

PRIMATOLOGY: OVERVIEW OF THE PRIMATES

Why Study Primates?

Key Issues

- Why is the study of primates important to physical anthropology?

Primate Features

Key Issues

- What are the general features of primate species?

OVERVIEW OF THE PRIMATES

Prosimians

Features

Key Issues

- What features distinguish prosimians from anthropoids?

- What are the major prosimian groups?

Groups

New World Monkeys

Features

Groups

Key Issues

- What features distinguish New World Monkeys from Old World Monkeys?

- What are the major New World Monkey groups?

Old World Monkeys

Features

Groups

Key Issues

- What are the major Old World Monkey groups?

Apes

Features

Groups

Key Issues

- What features distinguish Old World Monkeys and Apes?

- What are the ape species?

PRIMATOLOGY: OVERVIEW OF THE PRIMATES

Primate Phylogenies Exercises

Fill in the phylogeny with the following ape species (listed in alphabetical order): Bonobo, Chimpanzee, Gibbon, Gorilla, Human, Orangutan

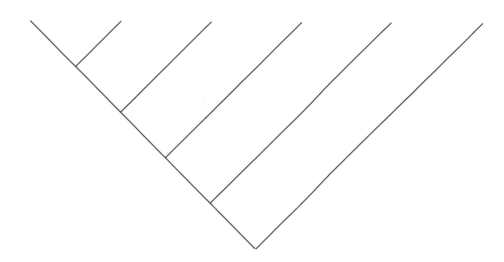

Fill in the phylogeny with the following primate species (listed in alphabetical order): Bonobo, Bush Baby (Prosimian), Chimpanzee, Gelada Baboon (Old World monkey), Human, Masked Titi Monkey (New World monkey)

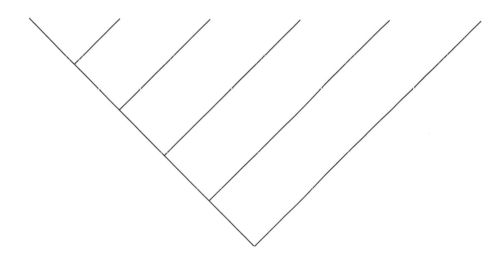

PRIMATOLOGY: ECOLOGY AND BEHAVIOR

BACKGROUND

The Language of Evolutionary Explanations

The Levels of Explanation

Ultimate explanations

Proximate explanations

Flexible Behavior

Soapberry bugs

Flexible behavior

The Level of Selection

Key Issues

- How do physical anthropologists describe behavior?

- How are proximate and ultimate explanations different?

- Can proximate and ultimate explanations coexist?

- Can flexible behaviors evolve?

- Can animals behave differently even if they have the same genes?

- At what level does natural selection act?

- What happens to traits that are bad for individuals but good for groups?

PRIMATOLOGY: ECOLOGY AND BEHAVIOR

PRIMATE ECOLOGY

Activity Patterns

Obtaining Resources

Size and diet

Quality of food

Difficulties

Territoriality

Ranges and territories

Costs and benefits / Patterns

Predation

Emigration

PRIMATOLOGY: ECOLOGY AND BEHAVIOR

GROUP LIVING

Costs

Resource Defense Model

Predation Avoidance Model

Key Issues
• What are the costs of living in groups?
• What is the basis of the Resource Defense Model?
• What is the basis of the Predation Avoidance Model?
• What evidence supports and what evidence goes against each of the models of group living?

GROUP LIVING EXERCISES

You are an anthropology graduate student studying group living and need to convince your graduate advisor that you understand the complexities of the topic. Evaluate the data below.

The chart below shows the relationship between group size and the percentage of encounters won between groups. Does this data support or contradict the Resource Defense Model?

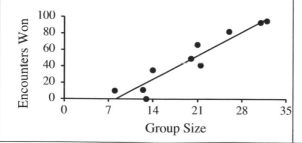

The chart below shows the relationship between capuchin monkey group size and predator scanning rate. Does this data support or contradict the Predation Avoidance Model?

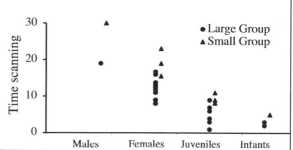

Black–and–white colobus monkeys live in cohesive social groups and rely mainly on leaves. Would this data support or contradict the Resource Defense Model? In other words, should primates group together to defend access to leaves? Why or why not?

Compared to larger groups of vervet monkeys, small groups have more incursions into their ranges and defend their territories more aggressively. How does this evidence impact the Resource Defense Model?

Juvenile monkeys suffer higher mortality in small groups than in large groups in habitats where predators are present. Does this support or contradict the Predation Avoidance Model?

PRIMATOLOGY: ECOLOGY AND BEHAVIOR

ALTRUISM

Altruism is a Puzzle

Types of social interactions

The puzzle

Kin Selection

Overview

Reciprocal Altruism

Overview

Requirements

Key Issues

- Why is altruism a puzzle?
- Why cannot purely altruistic acts evolve?
- What is kin selection?
- Why is kin selection not purely altruistic?
- What is reciprocal altruism?
- Why is reciprocal altruism not purely altruistic?
- What are the underlying conditions necessary for reciprocal altruism to occur?
- Why is kin selection observed more often?

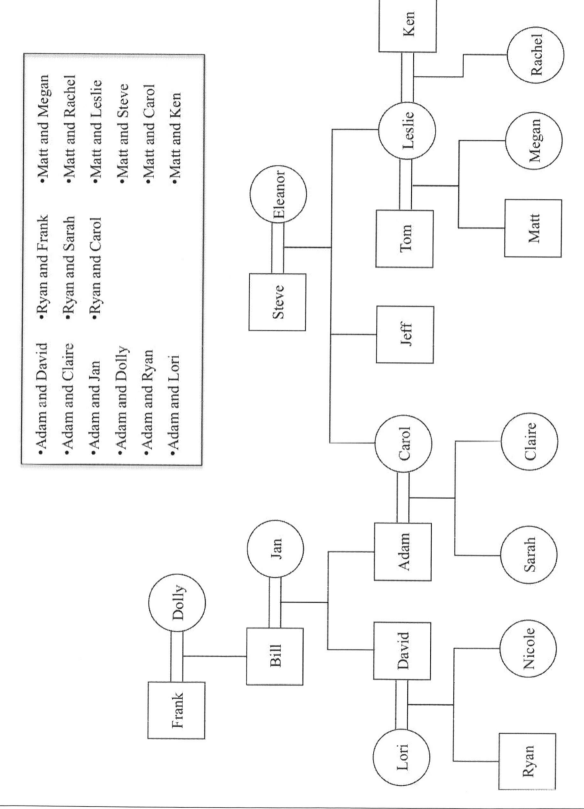

- Adam and David
- Adam and Claire
- Adam and Jan
- Adam and Dolly
- Adam and Ryan
- Adam and Lori

- Ryan and Frank
- Ryan and Sarah
- Ryan and Carol

- Matt and Megan
- Matt and Rachel
- Matt and Leslie
- Matt and Steve
- Matt and Carol
- Matt and Ken

PRIMATOLOGY: ECOLOGY AND BEHAVIOR

PRIMATE INTELLIGENCE

Background

Key Issues
• Why is studying primate intelligence important?
• What proposals have been made to explain why primates are so smart?
• How are the hypotheses for primate intelligence tested?
• To what extent to primates have a theory of mind?

Hypotheses for Primate Intelligence

Patchy–Resource Exploitation Hypothesis

Extractive Foraging Hypothesis

Social Intelligence Hypothesis

Evidence

Neocortex ratio

Scientific

Deception anecdotes

Limits of Primate Intelligence

Theory of Mind

The "Candy" game

PRIMATOLOGY: PRIMATE PARENTING AND MATING

INVESTMENT IN OFFSPRING

Variation in care

Factors influencing care

> **Key Issues**
>
> - What factors influence which parent invests in offspring?
> - Why do female primates typically invest more in offspring than males?
> - Why do both sexes in birds typically care for offspring?

1. Reproductive physiology / Mating dynamics
2. Whether additional matings lead to increased reproductive success
3. Whether care by two parents is better than care by one

Example: Birds and primates

MATING SYSTEMS

Solitary

Monogamy

Single–Male Polygyny

Multi–Male Polygyny

Polyandry

> **Key Issues**
>
> - What are the different types of mating systems?
> - What are the mating systems of each of the ape species?
> - What are the size differences between males and females in each of the ape species?
> - What types of ranges and/or territories do each of the ape species maintain?

PRIMATOLOGY: PRIMATE PARENTING AND MATING

SEXUAL SELECTION

Overview

<table>
<tr><td>

Key Issues

- What is the relationship between parenting and mating?

- What did Bateman's research reveal about reproductive success?

- What is sexual selection?

- What is the difference between intra- and inter-sexual selection?

- What are typical male and female sexually-selected traits?

- What is a sexually-reversed species?

</td></tr>
</table>

Bateman's principle

Sexual selection

Forms

Typical traits

PRIMATOLOGY: PRIMATE PARENTING AND MATING

REPRODUCTIVE STRATEGIES

Female Reproductive Strategies

Investment

Dominance hierarchies

Key Issues
• How can female primates increase their reproductive success?
• What is a dominance hierarchy?

Male Reproductive Strategies

Morphological traits

Behavioral traits

Key Issues
• What is sexual dimorphism?
• What is the relationship between primate mating systems and sexual dimorphism in body and testes size?
• Under what conditions are male primates likely to engage in infanticide?

Primate Sexuality Exercises

Listed below are primate "Personal–Ads" for partners. For each, indicate the mating system of the individual(s) doing the solicitations.

1. Several female monkeys interested in finding a few good males to mate with. Parental care not required.	5. Orangutan male looking for a good time. Only in town for a while, so expect a one–night stand.
2. Group of sexy bonobo swingers looking for matings of all types. Age is not an issue. Same sex matings ok.	6. Huge male silverback gorilla looking for a few good females. High levels of sex not required.
3. Need protection from infanticidal males? Look no further. I am a young, robust male monkey waiting to take over.	7. Old World monkey male with **large** testicles interested in meeting females with sexual swellings.
4. Two stable tamarin brothers looking for a mate. Will be willing to carry offspring. Love twins.	8. Swinging gibbon female looking to settle down and raise a family. Willing to protect a territory.

Sexual Selection Exercises

Listed below are descriptions of various traits for several different species. Answer each of the questions given below.

1. Phalaropes are shorebirds found in North America. At their breeding grounds, males and females pair off, but once the eggs are laid, parental care is done solely by the males. (a) Is this a typical species or a sexually-reversed species? How do you know? (b) Which sex is more likely to engage in aggressive encounters? Which sex is more likely to be the choosier sex?

2. In a species of guppies, females prefer males with more orange color. (a) How can you determine if this is a typical or sexually-reversed species? (b) What are the typical male and female traits [such as investment in offspring and others]? (c) Is coloration the result of inter- or intra-sexual selection?

3. Male sticklebacks (fish) choose heavier females as mates. (a) How can you determine if this is a typical or sexually-reversed species? (b) What are the typical male and female traits [such as investment in offspring and others]? (c) Is female body size the result of inter- or intra-sexual selection?

PRIMATOLOGY: MIDTERM 2 OVERVIEW

GENERAL INFORMATION

The midterm is worth 150 points and covers information presented in the lectures. You must bring a pencil and a Scantron Test form. A bluebook will not be required. There will be 25 objective questions (5 points each) and 2 longer problems (25 points combined). In addition, you must complete the 10-point online portion of the midterm by the date specified on Canvas. You will have the entire class time to complete the exam. There should be very little time pressure. Cheating will not be tolerated.

SAMPLE SHORT-ANSWER ESSAY PROBLEMS

Describe how the process of natural selection leads to evolutionary change. *Hint: See the Adaptation Exercises example on page 5.*

Construct a phylogeny of all of the apes and the other major primate groups.

In class you were given two models to explain why primates live in groups. Answer the following questions related to this topic in sentence format [only four or five sentences are required]: (a) What are the costs to primates for living in groups? (b) Name the two models and briefly describe what they propose. (c) During an in-class exercise, you were presented with several pieces of information which either supported or contradicted the models. Give one piece of supporting evidence for one of the models and state which model is supported by that evidence.

In class you were given three hypotheses for why primates are so smart. Answer the following questions related to this topic in sentence format [only four or five sentences are required]: (a) Describe each of the three hypotheses. (b) Which one receives the most support? (c) Give one piece of supporting evidence for the hypothesis receiving the most support.

In class you were given two explanations for why primates engage in behaviors that benefit others. Answer the following two questions relating to altruism: (a) Why is altruistic behavior not favored by natural selection? (b) Describe the two explanations for why primates help others.

Several issues relating to primate mating systems were discussed in class. In the space provided, answer the following questions: (a) Why do primate females typically spend more time investing in offspring than males? (b) What is sexual selection? Be sure to explain why males typically have structures for fighting or conspicuous traits that may make them more susceptible to predators. (c) Pick one trait (either physical or behavioral) related to primate mating and describe that trait from the perspective of sexual selection.

Species

What is a species according to the Biological Species Concept?

How do new species emerge?

What is a niche?

What is an adaptive radiation?

What is a phylogeny?

What are derived and convergent traits?

What are the differences between Cladistic and Evolutionary taxonomy?

PRIMATOLOGY: MIDTERM 2 OVERVIEW

Primates

Why are primates important to study?

1.

2.

Primate Features	
General Features	
Prosimians	
New World Monkeys	
Old World Monkeys	
Apes	

Ape Mating Systems	Bonobo
	Chimp
	Gorilla
	Orangutan
	Gibbon / Siamang

EARLY HOMININS: STUDYING THE PAST

The Fossil Record

Fossils

Taphonomy

Paleospecies

Incomplete record

Problems

> ## Key Issues
>
> - What is a fossil?
> - What is taphonomy?
> - What is a paleospecies and why do paleoanthropologists use this term?
> - Why is the fossil record incomplete?
> - What problems do an incomplete fossil record cause for paleoanthropologists?

Dating Methods

> ## Key Issues
>
> - What are the various dating techniques?
> - Under what conditions would a relative dating method be used?

Technique	Date Range	Materials	Relative / Absolute

EARLY HOMININS: STUDYING THE PAST

The Changing World

Formation of the Earth and Moon

Continental drift

World climates

Key Issues

- When was the Earth formed?
- What is continental drift?
- How does continental drift influence evolution?
- What has happened to the world's climate over the past 100 million years?

The Diversification of Life

Beginnings of life

Early life forms

Key Issues

- When did life first begin?
- When did multi–cellular animals first appear?

The Diversification of Plant Life

Gymnosperms and angiosperms

Key Issues

- What are gymnosperms and angiosperms and when did they first appear?
- Why are angiosperms important?

The Diversification of Animal Life

Early animal forms

Mammals

Key Issues

- When did mammals first appear?

Primate Evolution

Key Issues

- When did the various primate groups appear?

EARLY HOMININS: EMERGENCE OF THE HOMININS

Background

Apes and Hominins

Key Issues

- Which primates are in the ape and hominin groups?

- What features distinguish hominins from other apes?

- What are the components of species' names?

1.	Bipedal locomotion
2.	Increased brain size in relation to body size
3.	Features of dentition and jaw musculature
4.	Juvenile dependence / sophisticated culture

Hominin Evolution

Species' Names

Origins

Piltdown Man

Key Issues

- What was significant about the Piltdown specimen?

- When did humans and apes last share a common ancestor?

- When did hominins first appear?

Timeline

EARLY HOMININS: AUSTRALOPITHECINES AND PARANTHROPINES

Emergence

Australopithecus afarensis

Key Issues
- When and where did *A. afarensis* and *A. anamensis* live?

Australopithecus anamensis

Bipedalism

Overview

Key Issues
- What traits reveal that australopithecines were bipedal?

Evidence – Pelvis

Evidence – Footprints / Feet

Evidence – Angle of the knee

Evidence – Spine curvature

Evidence – Foramen magnum

Evidence – Tibia

Evidence – Humerus

Australopithecus afarensis	*Australopithecus anamensis*
Eastern Africa	Eastern Africa
4.0 – 3.0 mya	4.2 – 3.8 mya

EARLY HOMININS: AUSTRALOPITHECINES AND PARANTHROPINES

Southern Species

Key Issues
- When and where did *Australopithecus africanus* live?

Brain Size

Key Issues
- What was the brain size of the australopithecines?

Dentition

Overview

Key Issues
- What were the dental features of australopithecines?

Dental arcade

Canine size

Diastema

Cusps

Sexual Dimorphism

Key Issues
- How sexually dimorphic were australopithecines?

Australopithecus africanus

Southern Africa

3.0 – 2.2 mya

EARLY HOMININS: AUSTRALOPITHECINES AND PARANTHROPINES

Why Bipedalism?

It's Handy

It's Cool

It's Efficient

Key Issues

- What hypotheses have been presented to explain the origins of bipedal locomotion in hominins?

PARANTHROPINES

Overview

Timeline

Key Issues

- When did the paranthropines split from the australopithecines?

Species

Key Issues

- When and where did the various paranthropine species live?

Morphology

Cranial Features

Function

Brain Size

Key Issues

- What were the cranial features of paranthropines?

- What were the dental features of paranthropines?

- What was the function of the cranial traits of paranthropines?

- What was the brain size of paranthropines?

- What were the post-cranial features of paranthropines?

Post-Cranial Features

Paranthropus aethiopicus	*Paranthropus boisei*	*Paranthropus robustus*
Eastern Africa	Eastern Africa	Southern Africa
2.6 – 2.2 mya	2.2 – 1.3 mya	1.8 – 1.0 mya

EARLY HOMININS: AUSTRALOPITHECINES AND PARANTHROPINES

OTHER EARLY HOMININS

Background

Key Issues
• What general features are found with other early hominins?
• Why was Ardi a significant find?
• Why are some specimens considered possible hominins?

Similar Species

Review

Australopithecus garhi (2.5 mya, Ethiopia)

Australopithecus sediba (1.8 mya, South Africa)

Australopithecus bahrelghazali (3.6 mya, Chad)

Kenyathropus platyops (3.3 mya, Kenya)

Ardipithecines

Ardipithecus ramidus (4.4 mya, Ethiopia)

Ardipithecus kadabba (5.6 mya, Ethiopia)

Possible Hominins

Orrorin tugenensis (6 mya, Kenya)

Sahelanthropus tchadensis (6 or 7 mya, Chad)

EARLY HOMININS: BEHAVIOR

Early Hominin Behavior Exercises

Discuss the implications for understanding early hominin behavior for each of the following pieces of information.

Tools

A few ape species have been observed making tools. Orangutans sometimes modify sticks to obtain honey and insects from trees. Chimpanzees use sticks to extract termites from mounds and place nuts in between stones to crush them open. Despite such evidence, tools have not been found at early hominin sites (with one possible exception).

- Does this mean that early hominins were not making or using tools?
- If you believe they were, then why have none been found?

Social System

Even though the early hominins are no longer around, you can determine the social system of these creatures. Recall the relationship between sexual dimorphism and social systems in primates.

- What were the dimorphism levels of australopithecines and paranthropines?
- What was the social system of these early hominins?

"The First Family"

At Hadar in Eastern Africa, remains of a group of perhaps 9 adults and 4 juveniles have been discovered. They may have all died in a single, catastrophic event such as a flash flood.

- What does a collection of individuals indicate about social organization?

Swartkrans

At the South African site known as Swartkrans, there is an enormous collection of australopithecine remains including many juveniles accumulated over time.

- What can account for this massive accumulation?
- Where might one find large collections of animal bones?

Early Hominins: Behavior

Background

Models of behavior

Key Issues
• What groups provide models for understanding hominin behavior?
• What types of environments were early hominins occupying?

Environments

Resource Procurement

Determining diets

Key Issues
• How can hominin diets be determined?
• To what extent do modern foragers and chimpanzees rely on meat?
• How does seasonality affect patterns of meat eating?
• What evidence suggests that paranthropines ate meat?
• To what extent were early hominins obtaining food through hunting and/or scavenging?

Diets of early hominins

Meat eating

Social Organization

Background

Important sites

Tree living

Social system

Maturation rates

EARLY HOMININS: QUIZ 1 OVERVIEW

GENERAL INFORMATION

The quiz is worth 100 points and covers information presented in the lectures. You must bring a pencil and a Scantron Test form. A bluebook will not be required. There will be 25 objective questions worth 4 points each. In addition, you must complete the 10-point online portion of the midterm by the date specified on Canvas. You will have the entire class time to complete the exam. There should be very little time pressure. Cheating will not be tolerated.

If you have perfect attendance during this portion of the class, you will not have to answer 3 of the questions which will give you 12 points automatically.

You will have 45 minutes to complete the test. There should be very little time pressure. Cheating will not be tolerated. Note: There will be a lecture after the quiz and **attendance will be taken**.

STUDY TIPS

The exam will cover information in the Paleoanthropology Overview as well as information about australopithecines, paranthropines, species prior to australopithecines, and early hominin behavior. Make sure you thoroughly review this information.

Fill in the worksheets in this section of the Study Guide with as much information as possible.

Do the exercises in this section of the study guide to help familiarize yourself with dates and other significant events.

Try to write out species names, dates, and locations on a blank sheet of paper.

Study the drawings of the skulls as they will appear on the quiz.

The Fossil Record

What causes make the fossil record incomplete?	What problems result from an incomplete record?

What is a paleospecies?

Dating Methods

	Absolute/Relative	Date Range	Material
Radiocarbon			
Potassium-Argon			
Thermoluminescence			
Stratigraphy			
Fluorine Diffusion			

When did the following occur?

_____ Angiosperms emerge

_____ Apes appear

_____ First life forms

_____ Mammals appear

_____ Monkeys appear

_____ Multi-cellular organisms emerge

_____ Primates appear

Australopithecine Species

Species Name	Dates	Where Found?	
1.			
2.			
3.			
Brain Size / Dentition / Other Features			

Paranthropine Species

Species Name	Dates	Where Found?	
1.			
2.			
3.			
Brain Size / Dentition / Other Features			

Before Australopithecines

Group / Species	Dates	Location	Notes
Similar Species			
Ardipithecines			
Possible Hominins			

EARLY HOMININS: QUIZ 1 WORKSHEETS

Bipedalism

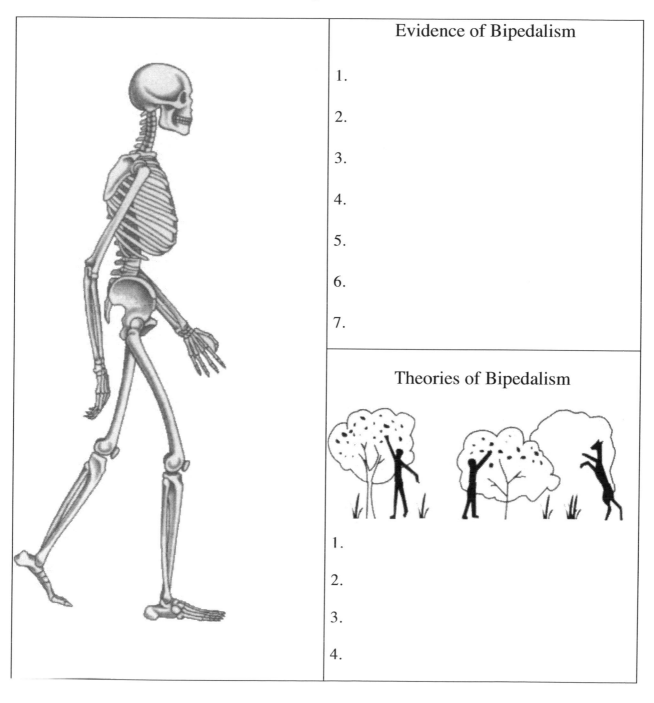

Evidence of Bipedalism

1.

2.

3.

4.

5.

6.

7.

Theories of Bipedalism

1.

2.

3.

4.

Early Hominin Behavior

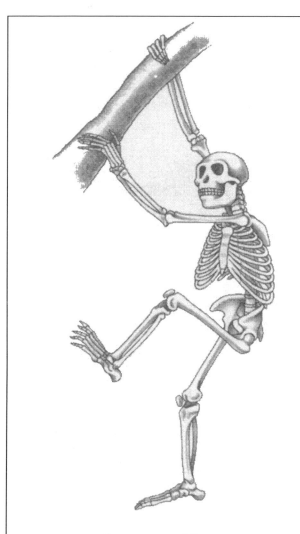

Important Sites

Hadar

Swartkrans

Resource Procurement

How can hominin diets be determined?
1.
2.
3.
4.

What foods were australopithecines eating?

What foods were paranthropines eating?

Meat eating / Hunting vs. scavenging

Social Organization

The vestibular system reveals that

Sexual dimorphism levels reveal that

Maturation levels reveal that

THE GENUS *HOMO*: EMERGENCE OF THE GENUS *HOMO*

Background

Hominin Features Review

	Apes	Early Hominins	Later Hominins
Bipedalism			
Biting			
Brains			
Behavior			

Key Issues
- What features distinguish the early and later hominins?
- Who were the ancestors of the later hominins?

Timeline

Species

Overview

Number of species

Key Issues
- When and where did *H. habilis* and *H. rudolfensis* live?
- What were the cranial features of the early members of the Genus *Homo*?
- Why is there uncertainty about the number of species of early members of the Genus *Homo*?

Small and Large Forms

Post-Cranial Features

Homo habilis	*Homo rudolfensis*
Eastern and Southern Africa	Eastern Africa
2.0 – 1.6 mya	2.4 – 1.6 mya

Oldowan Tools

Overview

Utility

Important Sites

Meat Eating

Evidence of hunting

Extractive Foraging

Significance

THE GENUS *HOMO*: GENUS *HOMO* DIVERSIFICATION

Overview

    Hominin Features

    Timeline

Homo ergaster

    Overview

    Cranial Features

    Post-Cranial Features

    Behavior

Homo ergaster
Africa and Eurasia
1.8 – 0.6 mya

Homo erectus

Overview

Morphology

Behavior

Homo heidelbergensis

Overview

Cranial features

Post-Cranial features

Behavior

Homo erectus	*Homo heidelbergensis*
Eastern Asia	Africa, Eastern Asia, and Western Eurasia
1.8 mya ~ 30 kya	800 kya ~ 100 kya

THE GENUS *HOMO*: GENUS *HOMO* DIVERSIFICATION

Homo neanderthalensis

Overview

Morphology – Cranial features

Morphology - Post-cranial features

Behavior

Other members of the Genus *Homo*

Homo floresiensis

Homo naledi

> ## Key Issues
> - When and where did *Homo neanderthalensis* live?
> - What are the important cranial and post-cranial features of *Homo neanderthalensis*?
> - What was the brain size of Neandertals?
> - What tools did *Homo neanderthalensis* make?
> - What evidence supports the fact that *Homo neanderthalensis* hunted large game animals?
> - What other important behavioral traits did Neandertals exhibit?

> ## Key Issues
> - When and where did other members of the Genus *Homo* live?
> - What were the significant physical features of other members of the Genus *Homo*?

Homo neanderthalensis	*Homo floresiensis*	*Homo naledi*
Europe and the Middle East	Flores Island, Indonesia	South Africa
130 kya ~ 30 kya	38 kya ~ 18 kya	250 kya

THE GENUS *HOMO*: *HOMO SAPIENS*

Overview

Timeline

Origins

Migration

Key Issues

• When and where did *Homo sapiens* emerge?

• When did *Homo sapiens* migrate to other parts of the world?

Morphology

Cranial features

Key Issues

• What are the distinguishing physical features of *Homo sapiens*?

Post-cranial features

Behavior

Background

Tools

Key Issues

• When and where did Upper Paleolithic behaviors emerge?

• What kinds of tools were Upper Paleolithic peoples using?

> *Homo sapiens*
>
> Worldwide
>
> Emerge 200 kya

THE GENUS *HOMO*: *HOMO SAPIENS*

Behavior (continued)

Food procurement

Symbolic behaviors

Environmental coping

Upper Paleolithic lifestyle

Language

Adaptation

Emergence

Key Issues

- What kinds of foods were Upper Paleolithic peoples eating?

- What kind of symbolic behaviors were expressed by Upper Paleolithic peoples?

- How were Upper Paleolithic peoples better at coping with their environments than Neandertals?

- What were the lives of Upper Paleolithic people like?

Key Issues

- What is meant by the statement "language is an adaptation"?

- What evidence supports the proposal that language is an adaptation?

- What is the evidence that language evolved relatively early in human evolution?

- What is the evidence that language evolved relatively late in human evolution?

THE GENUS *HOMO*: QUIZ 2 OVERVIEW

GENERAL INFORMATION

The quiz is worth 100 points and covers information presented in the lectures. You must bring a pencil and a Scantron Test form. A bluebook will not be required. There will be 25 objective questions worth 4 points each. In addition, you must complete the 10-point online portion of the midterm by the date specified on Canvas. You will have the entire class time to complete the exam. There should be very little time pressure. Cheating will not be tolerated.

If you have perfect attendance during this portion of the class, you will not have to answer 3 of the questions which will give you 12 points automatically.

You will have 45 minutes to complete the test. There should be very little time pressure. Cheating will not be tolerated. Note: There will be a lecture after the quiz and **attendance will be taken**.

STUDY TIPS

The quiz covers information presented in the lectures. The exam will cover information on the Genus *Homo*. Make sure you study this information.

Fill in the worksheets in this section of the Study Guide with as much information as possible.

Do the exercises in this section of the study guide to help familiarize yourself with dates and other significant events.

Try to write out species names, dates, and locations on a blank sheet of paper.

Study the drawings of the skulls as they will appear on the quiz.

Early Members of Genus *Homo*

Species Name	Dates	Where Found?	
1.			
2.			
Brain Size / Dentition / Other Features			

Oldowan Tools

The key differences between early hominins and later hominins are

  1.

  2.

When did Oldowan tools appear?

How useful were Oldowan tools?

Why was Olduvai Gorge important?

To what extent did early members of the Genus *Homo* eat meat and hunt?

What is the significance of Oldowan tools?

THE GENUS *HOMO*: QUIZ 2 WORKSHEETS

Homo ergaster / Homo erectus

Species Name	Dates	Where Found?	
1.			
2.			
Brain Size / Other Features / Tools Used / Behavior			

Homo heidelbergensis

Dates	Where Found?	
Brain Size / Other Features / Tools Used / Behavior		

Homo neanderthalensis

Dates	Where Found?	
Brain Size / Other Features / Tools Used / Behavior		

Other Members of the Genus *Homo*

What was significant about *Homo floresiensis* and *Homo naledi*?

Homo sapiens

Dates	Where Found?	
Brain Size / Dentition / Other Features		

Dates of Migration to	Significant Behavior
Americas	Tools Used
Asia	Symbolic Behavior
Australia	Food Procurement
Europe	Environmental Coping
Middle East	
Origin of Language	
Evidence that language is an adaptation	
Evidence that language evolved early	
Evidence that language evolved late	

MODERN HUMANS: EVOLUTION AND CREATIONISM

Overview

┌─────────────────────────────────────┐
│   Key Issues   │
│ ·······························   │
│ • What considerations are important │
│   when exploring issues relating to │
│   evolution and creationism?   │
└─────────────────────────────────────┘

Common Misconceptions

If someone believes in evolution, he or she cannot believe in God

┌─────────────────────────────────────┐
│   Key Issues   │
│ ·······························   │
│ • What   are   some   common │
│   misconceptions about evolution?   │
└─────────────────────────────────────┘

Evolution is only a theory and therefore we should not place much reliance on it being true

The chance of complex adaptations evolving is remotely small

The fossil record does not support evolution

Overview

Background

Sources of variation

Interactions

Nature / nurture debate

Patterns of Variation

Definitions

Adaptation - A trait designed by natural selection

Acclimatization Effect - A short-term physiological response to the environment

Responses to heat

Responses to cold

Patterns of Variation (continued)

High altitudes

Lactose intolerance

Skin color

Key Issues

• What acclimatization effects and adaptations are associated with high-altitude environments?

• What adaptation is associated with the ability to consume milk products?

• What are the advantages and disadvantages of having darker or lighter skin?

The Race Concept

Significance

Misconception #1 – People can be naturally divided into different races.

Key Issues

• From a biological perspective, what is wrong with the concept of race?

Misconception #2 – Members of each race are genetically similar to one another.

Misconception #3 – Members of different races are different in important ways.

PREGNANCY

Protecting Offspring

Morning sickness

Parent-Offspring Conflict

Maternal-Fetal Conflict

Spontaneous abortion

Blood sugar

GROWING UP

Conflict between siblings

Personalities of firstborn children and latterborn children

AVOIDING THREATS

Food Threats

Spices

MENOPAUSE

Background

Hypotheses to explain menopause

Dying

Background

Mutation Accumulation

Antagonistic Pleiotropy

Curing aging

MODERN HUMANS: EVOLUTION AND THE HUMAN MIND

BACKGROUND

General Principles

Genetic determinism

Naturalistic fallacy

Key Issues

- Why is applying evolutionary theory to human behavior controversial?

- What is "genetic determinism" and why is the idea incorrect?

- What is the "naturalistic fallacy"?

- What is evolutionary psychology?

- What are the major theoretical components of evolutionary psychology?

Genetic Determinism – This is the idea that only genes determine who we are at all levels: physical, emotional, and behavioral

Naturalistic Fallacy – This is the false assumption that what we see in nature is right, just, and inevitable. However, knowing how something is, tells us nothing about how that thing should be.

Evolutionary Psychology

Research program

Theory

COOPERATING WITH OTHERS

Review

Reasoning about Reciprocity

Survey

Interpretations

Key Issues

- What are the evolutionary influences that lead people to cooperate with each other?

- What do the results of the selection task tell us about the human mind?

COOPERATING WITH OTHERS

Economic Decision Making

Experiment

Results / Interpretations

Key Issues

- What do the results of the economic decision making tasks tell us about the human mind?

- How are people affected by the presence of others when making decisions about generosity?

Managing Reputations

Experiment

Results / Interpretations

HELPING RELATIVES

Review

Identifying Relatives

Paternity uncertainty / Parental resemblance

Attributions

Key Issues

- What are the evolutionary influences that lead people to help relatives?

- What patterns were found relating to how people attribute resemblance of relatives?

- What patterns were shown relating to support given by Hadza fathers?

- What patterns were found relating to grandparent support?

Paternal Support

Hadza study

Grandparent Support

Background

Results

MODERN HUMANS: EVOLUTION AND THE HUMAN MIND

The Human Mind Exercises

The questions below are based upon the following article about the book *Rape*.

1. Your brother is applying for a faculty position at UCLA and you are an audience member at his job talk. As part of his talk, he notes that Thornhill and Palmer argued that rape is an evolved predisposition designed by natural selection to enhance reproduction. A prominent member of the sociology faculty angrily notes that many social scientists suggest that the media is, in part, responsible for rape by bombarding the public with violence and sexually suggestive images. The sociologist then states that your brother's conclusions clearly must be wrong. You see that your brother is getting flustered and your task is to help him out. Using information gained in this class, what could you say to help your brother? Hint: return to the Levels of Explanation sub-topic in the Primate Behavior section and consider proximate and ultimate levels of explanation.

2. Congratulations! You have graduated from community college, gone on to receive your bachelor's degree, and successfully completed law school. You are working as a prosecuting attorney in Los Angeles when, in the midst of a sexual assault trial, you are hit with a bombshell. The defense attorney, relying upon the hypotheses of Thornhill and Palmer, notes that even if her client committed the act, that he should not be held responsible. The basis of this assertion is that "rape is in the genes". Your job is in jeopardy. How will you respond? Hint: think about the naturalistic fallacy as described recently in class.

3. You work with Randy Thornhill as a graduate student in anthropology. He has asked you to prepare a formal response to a claim contained in one of the letters to the editor. Specifically, the point was made that men sometimes rape 2-year-olds and 77-years-olds. What will your response be? Thornhill suggests that you should incorporate the notion of anecdotal evidence described in the Scientific Methodology section of the class.

4. Craig Palmer also needs to address an issue brought up in a letter to the editor. The writer pointed out that even in societies with high rates of rape, only a minority of men engages in the behavior. This appears to be counter to the suggestion that men may be "genetically predisposed to rape". Suppose you were a graduate student working with Craig Palmer and he assigned you to draft a response. What would you say? You will also need to address the question of whether or not there must be a rape gene which makes those men with the gene become rapists and those without non–rapists. You best option is to think about the discussions in the Human Variation section on nature versus nurture and the fact that genes are like recipes, not blueprints.

5. Suppose that after graduation, you started working at a governmental agency that makes recommendations on public policy. Your boss knows that you have taken a course in physical anthropology and has asked you to address two of the authors' claims relating to punishments for rape. She wants you to address two questions using an evolutionary perspective. First, can women provoke rape by dressing and acting in ways that might be sexually enticing, and, if true, should this be used as evidence for decreasing punishments for sexual assault? Second, if men are predisposed to rape, will increased punishment for such crimes reduce the incidence rates? You might want to consider the relationship between nature and nurture to answer these questions.

Article on and Letters of Response to *Rape*
by Thornhill and Palmer

(Miller, Martin. "Rape". *Los Angeles Times* 2 Feb. 2000: E1+
Letters to the Editor. *Los Angeles Times* 13 March 2000)

Everyone in this story agrees that rape is bad. It's morally repugnant. Its perpetrators should be imprisoned, and it demands eradication. But that's where the agreement ends and the arguments begin around an incendiary new hypothesis: All men are potential rapists. That's the unavoidable conclusion of "A Natural History of Rape: Biological Bases of Sexual Coercion," a controversial new book from MIT Press that has reopened the all-but-settled debate over the impetus for the crime. The authors--two academics in evolutionary biology and anthropology--argue that rape is primarily about sex and propagation, not about violence and humiliation, as has generally been accepted for 25 years.

While the prevailing view absolves the victim of any responsibility, the new book suggests women can, indeed, provoke rape and should take steps to prevent it, such as taking chaperons on dates and dressing demurely. And society shares a responsibility to inform men of their primitive sexual drives so they can better control them. "We were aware of the misguided criticisms that would rain down upon us," said Randy Thornhill of the University of New Mexico in Albuquerque, who co-wrote the book with Craig T. Palmer of the University of Colorado in Colorado Springs. "We knew people would see this as blaming the victim." Thornhill was right about that. Since their book was published this month, the scientists have been hit from all sides. Fellow scientists, legal scholars, sexual abuse counselors and feminists have slammed the book as a dangerous step backward in the war on rape. Some even fear the hypothesis could be used in a sort of "Darwin made me do it" legal defense for rapists.

Among the most vocal critics is Susan Brownmiller, whose 1975 book, "Against Our Wills: Men, Women and Rape," spearheaded reform of the nation's rape laws. Her work, which postulated that rape is about dominance and degradation, put the legal focus on the rapist instead of the victim's sexual past. "I haven't been this outraged in years. I nearly choked reading their book," said Brownmiller from her home in New York City. "These guys are desperate to mock feminist theory."

Thornhill, 55, remains undaunted by and even forgiving of the book's detractors. He believes the work has been misunderstood, in part because of political correctness, the public's lack of scientific sophistication and distorted coverage by the media. At stake, Thornhill said, is the possibility of significantly reducing or even eliminating rape. But first the truth, as he calls his view of rape, informed by evolutionary science, must come out. Until it does, current rape-prevention programs are doomed to fail because they identify only the environmental part of the problem.

Thornhill draws his theory from the animal kingdom. Evolutionary theory holds that the characteristics found in certain animals that allow them to survive to produce more offspring will eventually appear in every individual of the same species. Why? Because those animals will have more offspring. Those that don't, die out. "Crudely speaking, sex feels good because over evolutionary time the animals that liked having sex created more offspring than the animals that didn't," Thornhill writes.

Applying this principle to humans, Thornhill argues, means that natural selection over millions of years favored promiscuous males and

discriminating females. On the one hand, males need to invest little time and effort to reproduce. They can share in child-rearing but don't have to. Females require at least nine months to reproduce and additional months, even years, to breast-feed and raise the child. Because of these biological differences, females learned to carefully select their mates. But men still had to find a way to be selected. The most common methods through eons of time have been to demonstrate physical prowess and to gain wealth and status.

Here is where Thornhill departs from most evolutionary scientists when he argues that rape was a third strategy to reproduce. Males resorted to rape when they were "socially disenfranchised" or when they found women vulnerable to attack. Even the authors disagree about how deeply rape is embedded in the male psyche. Thornhill believes that rape is a more powerful force and is an "adaptive" behavior passed along in the genes. Palmer disagrees, arguing that rape is merely a byproduct of the other adaptations--namely, the strong male sex drive and male promiscuity. In either case, the authors agree rape has evolutionary origins. Thornhill notes that rape occurs in a variety of insects, birds, fish, reptiles, amphibians, marine mammals and nonhuman primates.

Rape data support their evolutionary hypotheses, the authors claim. First, most rapes are perpetrated upon women of child-bearing age. Next, in a vast majority of cases, rapists are only as violent as they need to be to subdue the victim. One study the authors cited found that rapists engaged in acts of additional violence--such as beating, slapping or choking--less than 22% of the time, thus not hampering their chances for reproductive success with the victim. And finally, the authors say, rape is an enormously traumatic psychological event, more so than other violent crimes. And among those raped, women of reproductive age suffer even greater psychological trauma, according to studies cited by the authors. These reactions make sense from an evolutionary standpoint, the authors argue. Rape threatens a

woman's reproductive interests by potentially robbing her of the opportunity to choose the best father for her offspring, they say.

The book has enjoyed some favorable reviews from the scientific community. University of Michigan professor of natural resources and environment Bobbi S. Low praised the book for contributing a "much-needed perspective and analysis to a topic that is emotionally charged."

But critics outnumber supporters. Evolutionary scientists were among the first to take aim. Evolutionary scientist Jerry Coyne of the University of Chicago said the authors fail to sufficiently explain why men rape old women, children and other men. Also, Coyne believes committing rape millions of years ago was probably a lot more difficult than it is today. Humans traveled in smaller groups, making women easier to protect. Also, women during childbearing years were usually pregnant or lactating. "You pick a woman [to rape] at random, and the chances that you'd have a baby were pretty remote," Coyne said.

David Buss, an evolutionary psychologist at the University of Texas at Austin, said that the book fails to show that rape has a special purpose from an evolutionary standpoint. He knocks the authors for drawing incorrect causal connections. "Human males use violence to achieve a variety of ends. They steal televisions. They steal food," he said. "But you wouldn't say, because men use violence to steal televisions that it's an evolved food-stealing strategy."

Other critics, most notably Brownmiller, blast the authors for claiming to be able to measure a woman's psychological response to rape. She questions the validity of the studies the book cites to demonstrate the depth of wrenching emotions a rape victim undergoes. "How do you quantify rape pain?" Brownmiller asked. "That whole section about guilt and pain is the weakest part of the book."

Many critics aren't as concerned about the science of the book as the social and legal

message it could convey. In an era when society has witnessed a steady stream of defendants slipping out of justice's grasp because of legal arguments about a variety of mitigating circumstances, some worry a creative rapist might argue he was at the mercy of his genes. "The concern is understandable," said Owen D. Jones, a professor at Arizona State University College of Law in Tempe. "Nothing is stopping anyone from raising an absurd defense. The question should be, what is the likelihood that it will succeed? None." Susan Estrich, a legal professor at USC and a rape victim, agreed. Just because there might be an explanation why someone commits a crime in no way excuses its commission, she said. "I would be horrified if the thesis of this book were accepted," said Estrich, who has written extensively on rape and the law. "But it's an irrelevant question legally."

If anything, some legal scholars believe that if the book's hypotheses are true, penalties for rape may actually grow more severe. Harsher punishments may further compel men to control their sexual urges, legal scholars say. Some attorneys are nevertheless troubled by the notion that the book could be used to lessen a rapist's sentence. "You never know with policymakers," said Marci Fukuroda, an attorney with the California Women's Law Center in Los Angeles specializing in violence against women. "They might take this thing way too seriously and push for some sort of mitigation. "Of course, if they did try to push it, there would be 10 times as many advocates to push the other way," she added.

For their part, Thornhill and Palmer, who dedicated their book to "the women and girls of our lives," have made it clear they support stricter punishments for rape. There is no excuse, they say. Men may be genetically predisposed to rape, but environmental factors--including education and a certainty of punishment--can overrule them. The authors don't dwell on the legal or moral fallout of their work. They see their mission as simply to disclose the facts. "As scientists who would like to see rape eradicated from human life,

we sincerely hope that truth will prevail," they write.

Letter #1 – I am disturbed by Martin Miller's Feb. 20 article on rape in which he reviews "A Natural History of Rape," by Randy Thornhill and Craig Palmer. As a psychologist who works with rape survivors, I decry what will be taken as justification for rape by men already looking for further excuses to act out their rages. The Thornhill-Palmer book is inadequately researched and reasoned. Further, I believe these men have no experience themselves with rape survivors. Theories are a poor substitute for practice. Miller notes that "women's advocates are outraged" by the book. In fact, all human beings should be outraged by distortions of fact that debase men as well as women. To say that rape is sexual (and therefore normal) because most victims are of childbearing age is like saying all men are violent because most violent crimes are committed by men. If rape is a genetic program to spread a man's seed, how do we explain the recent rapes of a 2-year-old and a 77-year-old? Both cases occurred in my neighborhood, as well as another involving the brutal sodomizing of a woman. Thornhill and Palmer say rapists usually use no more force than is necessary to complete the rape. This view is contradicted by hospital records of injuries, and trivializes the violence, both physical and psychological, by claiming it is sex and therefore not so bad. How do these men explain the high frequency of rape in prisons? No prison authority believes this is procreative behavior. It is domination. Even in rape-prone societies such as ours, only a minority of men rape. Saying how a woman dresses is a causal factor detracts attention from the characteristics of those felons who violate women (and men) against their will. To determine who rapes is a necessary step in holding rapists and society accountable.

MODERN HUMANS: PARENTING AND MATING

PARENTING

Adoption

Child abuse / Infanticide

Key Issues
• Why is adoption puzzling from an evolutionary perspective?
• How can human adoption be explained from an evolutionary perspective?
• Why is infanticide puzzling from an evolutionary perspective?
• How can patterns of infanticide be explained from an evolutionary perspective?

MATE CHOICE

Background

Inbreeding avoidance

Key Issues
• From an evolutionary perspective, why should individuals avoid mating with close relatives?
• What proximate mechanism prevents people from mating with close relatives?

MATE PREFERENCES

Attributes preferred

Patterns

Key Issues
• What attributes in partners do people prefer?
• Why do men and women place high values on the same attributes?
• Why do women prefer attributes relating to resources?
• Why do men prefer attributes relating to attractiveness?

THE LOOKS PARADOX

Overview

Women are choosier

Men are choosy

Traditional explanation

Wetsman's explanation

Key Issues

- Why are human males "unusual" with respect to the preferences they have for partners?

- What explanations were proposed for why men are "unusual" with respect to such preferences?

FINAL OVERVIEW

GENERAL INFORMATION

You must bring a pencil and a Scantron form. There will be 45 objective questions (3 points each) and 4 short-answer essays (worth 65 points combined). You will have 2 hours to complete the test. There should be very little time pressure. Cheating will not be tolerated. The exam will be cumulative and will cover information from the first three parts of the class. However, the majority of the multiple-choice questions will be on information covered after the second quiz starting with the topic of Evolution and Creationism.

ESSAY QUESTIONS

You will be given 4 short-essay questions to answer during the exam, worth 65 points combined. The essays and their point values are shown below. The first is the "Rest during the test" problem. If you have perfect attendance for the final section of the class, you will automatically receive full credit for that question.

- [10 Points] Why is race a meaningless concept from a biological perspective?

- [25 Points] During the last section of the class, several issues relating to evolution and human behavior were reviewed. A worksheet with several of the issues is provided on the next page of the Study Guide. **Choose one topic and answer the questions given in the Study Guide on the next page.** Possible topics include maternal-fetal conflict, birth order and personality, menopause, reasoning about reciprocity, economic decision-making, inbreeding avoidance, human mate choice, and male preferences for good looks.

- [10 Points] Explain how evolution by natural selection works.

- [20 Points] Suppose that you were at a party and someone asked you to briefly explain what happened in human evolution during the past 4 million years. What would you say? In other words, explain the sequence of events that changed an ape-like creature into a modern human. When and where did this transition take place? What were the significant features of the intermediate creatures? **Note:** Your grade for this question will be largely based upon how you summarize the information in layperson's terms. Again, you should answer as if you were talking to a person that knows little about the fossil record. Using phrases like "intermediate dental arcade" would not be appropriate in such circumstances.

GETTING YOUR GRADE

Grades will be posted on Canvas a few days after you take your final.

FINAL OVERVIEW: FINAL REVIEW WORKSHEETS

MATERNAL-FETAL CONFLICT

Why does maternal-fetal conflict arise? Be sure to explain how the differences between genetic relatedness contribute to conflict. What is the rate of spontaneous abortion? How do the mother and fetus attempt to influence the system of maintaining pregnancy? How are levels of blood sugar manipulated by the developing fetus?

BIRTH ORDER AND PERSONALITY

From a biological perspective, why is birth order expected to affect personality development? What are the different strategies used by firstborns and latterborns to solicit parental attention? What are the trends observed in the personalities of firstborns and latterborns? Try to give specific examples of famous people.

MENOPAUSE

From an evolutionary perspective, why is menopause unusual? Include information about other species. Give an evolutionary explanation for why menopause occurs in humans. What is the proximate cause of menopause in humans? Give specific figures.

REASONING ABOUT RECIPROCITY

What is the underlying logic of evolutionary psychology? Include information on special purpose mechanisms and what kinds of mechanisms are expected to evolve. Discuss the results of the psychological reasoning problems. Why is a mechanism for detecting cheaters important?

ECONOMIC DECISION-MAKING

How do humans act "economically irrational"? Why is this unusual from the perspective of evolutionary theory? What explanation can account for this behavior? Discuss the results of the "economic games". Provide some detail.

INBREEDING AVOIDANCE

From an evolutionary perspective, why is having children with close relative problematic? Be sure to include details about why the problem occurs. What mechanism has evolved in humans to prevent them from having offspring with relatives? Provide a specific example that was discussed in lecture or the textbook.

HUMAN MATE CHOICE

From an evolutionary perspective, why are males expected to favor some traits and females others? What are those traits? What traits do men and women find most important? What traits do men favor more? What traits do women favor more? Be sure to explain why these preferences make sense from an evolutionary perspective.

MALE PREFERENCE FOR LOOKS

Why are males "weird" from an evolutionary perspective with respect to their interest in physical attributes of partners? Compare men to males of other species. What explanations were given to explain human male preferences? Which sex is the choosier sex? To what extent are males choosy?

PAPER ASSIGNMENTS: PAPER ASSIGNMENT 1

GENERAL OVERVIEW

A project consisting of research at a zoo and some fact-finding will be worth twenty percent of the final grade. This project will be due in two parts as outlined below. Late papers will have their scores reduced as specified in the course syllabus. The end result will be a paper of approximately 2,000 words in length that describes the results of the research. Although the emphasis of the paper will be a description of the fact-finding conducted, credit will be given (and taken away) for style and grammar. Papers will be due at the beginning of class or online by a specific deadline

PART 1 DESCRIPTION

This part of the paper is worth eighty points, or eight percent of your final grade. The length of this part of the project should be about 1,000 words.

The purpose of this part of the project is to familiarize you with how primate research is done. In order to complete the task, you will have to go to a zoo (the Los Angeles zoo is great, but others are fine as well). You will choose two different **primate** species which you will observe for fifty minutes each. Selecting one ape species and one other species is a good way to see a contrast between primate groups, although this is not required. You will be observing how each allocates his or her time.

Choose **one** animal from each species. Then every minute, determine what that animal is doing. This will be recorded on your Time Allocation Sheet (available on the class website). For each species, you will have allocated 50 tasks (1 task per minute x 50 minutes). Examples of possible behaviors include eating, sleeping, grooming, aggression, communicating, playing, etc. You can also keep track of the number of other primates that are near your target animal.

When you write your paper, you should state which species you chose, the sex of the animals observed, if possible, and how the subjects spent their time. Make reference to the percentages of time allocated to each type of behavior. You should also compare and contrast the behavior of the two species observed. Finally, include a brief statement about how you felt about doing the project and what you learned. Do not worry about including the Time Allocation Sheets with your paper submission since they are just working documents to help you.

PAPER ASSIGNMENTS: PAPER ASSIGNMENT 1

MODEL EXCERPTS FOR PAPER 1

> Note: This was written by the instructor to provide an example of elements of a very good student paper.

Last Thursday, I went to the Los Angeles Zoo for the purposes of observing two primate species. I arrived at around eleven in the morning and decided to start my observations immediately since the day was rather warm and the sun was already getting hot. My first stop was at the new chimpanzee exhibit. The enclosure was quite impressive, having been recently constructed. There were about fifteen chimpanzees in the exhibit, which probably covered several acres. A waterfall was placed in the middle of the enclosure. The chimpanzees appeared to like this feature since many sat near the cascading water, probably to keep cool.

> Notice that the paper describes the events of the observation as well as the exhibit where the animals were being observed.

I decided to observe a large male chimp who I named Bill for the purposes of this description. I chose this individual since he was sitting near the glass display and I could get rather close to him initially. My superior vantage point did not last long (only about 4 minutes) because soon after I arrived the zookeepers put some fruit out near the other side of the enclosure. This resulted in some aggressive encounters between Bill and several other chimps. The aggression only lasted as long as the fruit, which was about four minutes in duration. Of the time spent observing the chimps, aggression accounted for 8% of their time. After the encounter, Bill spent some time grooming a smaller male chimp. This, and another grooming bout later during the observation, resulted in 14% of Bill's time being spent grooming.

> A detailed description of every action is not necessary, however, you may want to include noteworthy events in your paper.

> Include the percentages you calculated from your Time Allocation Sheet. Also include the sheets themselves. Base your percentages on your Time Allocation Sheets. DO NOT GUESS.

…The behavior of the chimpanzees was somewhat different from that of the Gelada baboons. First, the chimpanzee I observed spent more time in aggressive encounters, 8% compared to 0% for the baboon. However, this may be due to the fact that the chimpanzees were fed during my observation while the baboons were not. The baboons spent less time in close proximity to one another. Bill, the chimp, had an average of one other individual near him during my observation. This was much less for the baboon.

> Compare and contrast the behavior and time allocations of the two species. Include some details about what you observed.

…Although I understand the need for primate research, I found the time spent at the zoo pretty dull. Fifty minutes for each species was about all I could handle between the heat, monotony, and crowds. I have gained a great deal of respect for people who dedicate significant portions of their lives to the study of primates. I am glad I did the project, however, I would not relish the prospect of doing this again any time soon.

> Be sure to include a brief statement about your thoughts on this assignment.

Paper Assignments: Paper Assignment 1

Paper 1 Helpful Hints

Do the Time Allocation Properly – Remember that you should pick a single animal from two different species and observe each for 50 minutes. Do not merely estimate times spent doing each task. Also, do not keep track of more than one animal in each group. You <u>must</u> include the time allocation percentages in your paper. To calculate the percentages, simply count the number of times out of the fifty minutes that you observed a particular behavior and multiply by two. For example, if a primate was playing for fourteen out of fifty observations, the animal spent twenty–eight percent of the time playing. Provide exact numbers based upon your observations.

Start Early – Human nature often leads people to procrastinate. Getting caught the night the assignment is due trying to put the paper together at the last minute is a poor strategy for getting a good grade. The earlier you start, the better you will probably do.

Proofread, Proofread, Proofread – One of the most significant factors leading to reduced grades is when there are typos and spelling errors in a paper. This shows that the paper was put together in a rush. Read your paper before turning the final product in. Having a friend read your paper is a good idea as they will be more likely to catch errors.

Grammar Tips – Species names can be either lower or upper case (except at the beginning of a sentences), but should be consistent throughout the paper. Also, unless you are confident in their use, avoid semicolons. Another tip is to avoid abbreviating words. For example, use "did not" instead of "didn't". Finally, the plural of the word "species" is "species". You would never say, "I observed that the baboon specie…".

Monkeys, Apes, and Prosimians – Make sure that you reference your primates properly. Chimps, gorillas, orangutans, gibbons, and siamangs are apes. Lemurs and sifakas are prosimians. Referring to an ape or a prosimian as a monkey shows you were not paying attention in class and will result in lowered paper scores.

Appearance and Canvas Formatting – You will be turning in your paper via Canvas, which may do some interesting things with formatting. Do not worry about this too much. The content of the paper is more important than the appearance.

Use the Resources on the Next Page – The worksheet, checklist, and grading standards on the following page were developed to help you be successful. Use these resources!

Save Your Paper – While your submission should be stored in Canvas, it is a good idea to keep a copy somewhere else as well. The second part of the paper will be based on the first, so you will need it again.

PAPER ASSIGNMENTS: PAPER ASSIGNMENT 1

WORKSHEET

	Primate A	**Primate B**
Sleep / Rest	__ acts x 2 = ___%	__ acts x 2 = ___%
Eat	__ acts x 2 = ___%	__ acts x 2 = ___%
Aggression	__ acts x 2 = ___%	__ acts x 2 = ___%
Mate	__ acts x 2 = ___%	__ acts x 2 = ___%
Grooming	__ acts x 2 = ___%	__ acts x 2 = ___%
Other	__ acts x 2 = ___%	__ acts x 2 = ___%
Total	__ acts x 2 = <u>100</u>%	__ acts x 2 = <u>100</u>%

CHECKLIST

____ I read the "Helpful Hints" for this part of the paper included above.
____ I calculated the time spent for each activity using the worksheet above.
____ I incorporated these exact allocation times into my paper.
____ I provided a general introduction to my paper.
____ I described the two primates' enclosures and described their behaviors.
____ I compared the behaviors of the two primates in my paper.
____ I did not refer to any apes or prosimians as "monkeys".
____ I included some thoughts about the project into my paper.
____ I thoroughly reviewed my paper for grammar.

STANDARDS

There is one major element of this part of the paper: the description of the behavior of two primates using the time allocations determined from your research. There are also several minor elements: (a) a brief introduction, (b) a comparison of the species, and (c) a statement about what you thought about the project. Below is a rough guide for the level of work expected to obtain various grades. Generally all of the elements in a particular group must be present to obtain the given grade.

80 Points	70 Points
• A thorough description complete with allocations	• Incomplete or poor time allocations
• Complete incorporation of all minor elements	• A minor element missing or a few incomplete
• Few, if any, grammatical errors	• Few, if any, grammatical errors
60 Points	**50 Points or lower**
• Time allocations missing and/or poor description	• No time allocations and poor description
• One or more missing minor elements	• A few minor elements missing or incomplete
• Average grammar	• Poor grammar or other omissions

PAPER ASSIGNMENTS: STUDENT PAPER 1 SAMPLES

PAPER 1 – STUDENT SAMPLE A

On March 24 I attended the Los Angeles Zoo field trip to observe primates. When I arrived it was around ten in the morning and I went on the group tour. Our guide showed us where all the primates were located and gave us a brief description of each of them. The tour was helpful because we did not have to waste time figuring out where all the primates were. After the tour of the primate exhibits I decided to observe the gorillas. The exhibit was nicely put together. There were two levels to the exhibit. On the first level there was a small pond that they drank out of. There were also logs and rocks. There were small caves that provided shade for the gorillas. On the second level it was a grassy pasture that they seemed to enjoy laying down in. Inside the exhibit there were three gorillas. One of the gorillas was a male and the other two were female. I decided to observe one of the female gorillas and I will call her Sally.

I chose Sally because she was fairly easy to locate. When I first started the observation it was around eleven thirty and she was on the second level of the habitat. Since she was on the second level it was easy to observe her. She only stayed on the second level for about five minutes and then move down to the bottom level of the structure. Sally was eating the food that was off of the ground. The food looked similar to Cheerios. While I was observing Sally 46% of the time she was eating. She seemed to be moving around to find food on the ground.

When it was the ninth minute she was about three yards away from the other female gorilla. There were not many interactions between the three gorillas. They all seemed to be doing their own thing. In the twenty-first minute in which I observed Sally, she was about three yards from the male gorilla. The male gorilla was enormous in comparison to Sally. During this minute Sally seemed a little hesitant to do anything. She seemed intimidated by the male gorilla. After her encounter with the male gorilla she bent down to drink some water out of the pond. The way her body moved was very similar to the human body. Sally met with the other female gorilla two other times. In the twenty-eighth minute the other female gorilla was about three yards away from Sally while she was resting. A few minutes later the other female gorilla came back towards Sally and was within one to three yards for three minutes. Within the last minute they were playing with each other, but that was the only time I observed any kind of play. The playtime only took up 2% of the observation time.

Much of the rest of the time was spent resting or moving around for food. Resting took up 38% of the observation time. While Sally rested she was folding her arms much like we do. Sally would also yawn just like a human. She was yawning through out the observation, which led me to believe that she was getting tired and was about to go to sleep. In the last eight minutes of my observation of Sally she was resting and readjusting herself. Sally seemed to be getting ready for bed.

After I finished my observation of Sally I made my way over to the Geoffroy's spider monkey exhibit. By this time it was around twelve thirty and I was getting a little impatient. In this exhibit the spider monkeys were enclosed in a very large cage. The cage was filled with branches hanging from the ceiling of the cage. It was also filled with lots of tropical looking plants and was very shady. In the cage there was about seven spider monkeys. In the cage there was a small stream that the monkeys would get their water from. At this stream is where I found the spider monkey that I was going to observe.
The spider monkey that I observed was younger than the rest of the monkeys in the cage. The spider monkey was a male and I will name him Sal. After he got his drink of water he started playing.

Playing took up 34% of the observation time. Much of the playing that I observed was done in front of the heat lamp. While he was playing he was within one yard of either one or two monkeys. In the first fifteen minutes Sal spent most of his time playing. There was a lot interaction between the monkeys. They seemed very playful. During some of the playing it seemed that they were giving each other affection. Since he seemed to be the youngest spider monkey the others gave him a lot of attention. The affection took up 8% of the observation time.

Some of the play led to swinging on the branches in the cage. Swinging took up 18% of the observation time. When they would swing on the branches they would use their prehensile tails to help them reach from branch to branch. Sometimes Sal would hang by his prehensile tail. Hanging took up 4% of the observation time. At the twenty-eighth minute Sal stared grooming himself. Grooming took up 2% of the observation time. I also observed Sal eating something that looked like a nut. He held it in his hands holding it tightly and taking small, fast bites. Sal's eating took up 10% of the observation time. In between all the activity Sal would rest. Resting had taken up 30% of the time during observation. Most of the resting occurred within a yard of another monkey.

In observing the gorillas and the spider monkeys I noticed many differences. The spider monkeys were much more active then the gorillas. Sal spent 34% of the time playing while Sally spent 2% of the time playing. Sal was also closer to other monkeys for about 60% of the time. Only 14% of the time Sally was close to another gorilla. Among the monkeys there was more affection as well. Sally spent much more time eating then Sal. Both Sally and Sal spent 2% of the time grooming themselves.

My time spent at the zoo was very informative. Observing the spider monkeys using their prehensile tails was definitely interesting. It was amazing how it worked so well with the rest of their bodies. Observing the spider monkeys was relaxing because it was empty and peaceful. It was very crowded around the gorillas and I did not enjoy that as much. It was also a very warm day and the heat was annoying. But there is so much to learn from these animals.

PAPER 1 – STUDENT SAMPLE B

Recently in my Anthropology 101 class, we to a field trip to the Los Angeles Zoo. Our objective was to observe two primate for an hour and compare and contrasts the two. We took a tour with a Docent guide, who showed us all the primates at the L.A. Zoo. (Tenty-two to be exact) I had never really knew what a primate was or at least in great detail what it was, until I took this class and the field trip. The zoo was incredible, so many different kinds of animals; especially "primates". The two primates that interested me the most were the gorillas and the chimpanzees. The more we talked and learned about them in class, the more I wanted to see them in action. And well I did.
Many of us have heard about the mythical characters "Tarzan". Based on the myth, Tarzan was abandoned in the jungle by his parents who were killed by wild animals. Tarzan was found by a family of gorillas. In which they accepted him as another animal in the jungle. They taught him how to live and survive in the wild, to make do with what he had. I know this sound crazy to think that an actual gorilla has the ability to teach a human how to live. Gorillas are very much like humans, both having the natural instincts of right and wrong, values and family pride; just like us humans. Those are just a few examples of their personality that interested me in observing them. As our Docent guide led us around to the front of the Gorilla exhibit, I instantly made eye contact with the biggest creature that I had ever seen. I was so intrigued by this animal that I decide to observe him first. The long await was about to begin.

Since I was going to be watching him for an hour, I figured that it was only right to give him a name... "Preston". Preston and I sat for about six minutes just starring at each other. In away it was kind of uncomfortable; as if he was letting me know how it feels to have people staring at you all the time. Preston wasn't one of the most active gorillas among the three. About 40% of my his and my time were resting. Literally, he didn't budge! Ten minutes into my observation he picked some grass and ate it. It must have given him just a enough energy to keep him active for a couple o f minutes or until his arms got tired of all the pressure being on them. Even though Preston sat and rested the majority of the day, the next best thing to resting was trying to stay cool and out of the sun. Preston walked from one end of his enclosure to the other. (24% movement) When my observing time with Preston was ending. I got up to pack everything together, when out of no where it seemed as if this violent/aggressive behavior ran through out his body. I waited until he calmed down before I left. I 'm sure it was just a coincidence, but as I walked away he went and sat down under a little cave formed by rocks. He sat there with his back facing the crowd starring at the rocks. Not even paying attention to te peope. Just resting nothing new to him!

"Resting" was something that the baby Chimps just didn't know anything about. The second primate that I observed were the baby chimpanzees, one imparticular named "Jake". Jake was a very animated chimp that never really slowed down. He was constantly climbing on ropes, pushing objects out of his ways, swinging on hanging tires, etc. Jake spent 50% of his time while I was there, moving around and playing. The exhibit was filled with different toys that the chimps were able to play with. Everything was a hands on object where they can learn how to do things on their own. For example, they have ropes that are hung from the ceiling where they can get used to that kind of environment. One of the things that interested me the most was that one of the walls were painted with a jungle scenery. I thought that it might have been for them to get to the idea that that's what reality really is for them, not being locked up at a zoo. There were three chimps in one exhibit, they all did their own thing but at the same time they all were very close to each other. Even though only .04% was spent on grooming each other or themselves, they managed to always pick at each other. Aggression never really broke out between the three, with the exhibit being small, the chimps were always within one to three yards of each other. Never really giving them the opportunity to posses or take control of something. The chimps were a joy to watch, it kinds put a smile on your face. They were all full of so much energy just like children. (Just a little-bit more hairy) One of the members of the crowd managed to get Jakes attention and have him go over to the glass window where everyone was. The chimp just looked at everyone as if we were the one that were locked up; not them. He sat there watching for about four minutes and decide that food was a little more important than watching us. Though before he got up to leave he pressed his lips against the glass window. Which of course made everyone's heart melt; including mind.

The behavior between the two primates were in away very similar to each other. Preston was very human like. His eyes seemed as if they could tell you his life story, the fear, the anger, and the passion. With Jake he's just full of life, ready to go out there and create a story of his own. Both gorillas and chimps play a big role in our life and our society. One by giving us the use of our body and mind, and second by giving us the knowledge and "know how" abilities to live and survive in an everyday life. Without the basic skills of these primates, I don't think we'd get to far in the world.

Grammar
(Spelling, Punctuation, Sentence Formation, Paragraph Formation)

__ Poor   ___ You need to work on word choice, spelling and/or punctuation.
__ Acceptable   ___ You need to work on sentence structure.
__ Average   ___ You need to work on paragraph structure.
__ Very Good   ___ You need to work on verb tenses.
__ Excellent   ___ You should visit the Writers' Resource Center for help.

Overall Description

__ Poor   ___ You need to provide more detail in your paper.
__ Acceptable
__ Average   ___ Add more information about the primates' enclosures and habitat.
__ Very Good   ___ Put in more information about interesting events that occurred.
__ Excellent   ___ Apes and prosimians are not monkeys.

Time Allocation and Integration

__ Poor   ___ You need to incorporate time allocations into your paper more.
__ Acceptable   ___ You observed the group, instead of an individual primate.
__ Average
__ Very Good   ___ Check your time allocations. A calculation error has been made.
__ Excellent   ___ You were supposed to identify only one act per minute.

Other Elements

__ Poor
__ Acceptable   ___ You need to put more into your species comparison.
__ Average   ___ You need to put more into your thoughts about the project.
__ Very Good   ___ The length of your paper is too short. Add more detail.
__ Excellent

Points assigned to the paper ____

PAPER ASSIGNMENTS: PAPER 2 ASSIGNMENT

PART 2 DESCRIPTION

This part of the paper is worth 120 points. The purpose of this part of the project is to introduce you to library and/or Internet fact-finding. Using library and/or Internet sources, you should investigate the primates you observed for the first part of your project. You should find out where the primates observed live, their group size, types of food eaten, their population status (whether or not they are endangered), and any other information you deem relevant about the species. This should be done for both species observed in Part 1.

Incorporate the information you obtained into your first paper. Also include a few observations about how you felt about doing this part of the project like you did in the first part. The second paper you turn in should be about 2,000 words in length.

A key difference between average papers and very good papers is how well the various components are integrated. Students who put the fact-finding and the zoo observations in separate sections and who make little or no reference to their time allocations will not do so well. Those who weave the discovered facts in with what they saw at the zoo along with the calculated behavioral percentages will do much better.

PLAGIARISM / REFERENCING THE OUTSIDE SOURCES OF INFORMATION

Plagiarism is using another person's ideas without acknowledgment. Outside information is necessary for you to complete your paper. The only step you need to take is to indicate where you obtained the information, both within the body of your paper and in a bibliography. Examples can be found in the sample papers. If the information comes directly from the source, you must use quotation marks. Changing a word here or there will not suffice. There is nothing wrong with using direct quotes, however, try to avoid a lot of lengthy ones in your paper.

You should include a bibliography as well. Using MLA style or any other is fine so long as you are consistent and include all of the relevant information about your source. There are some great resources available online to help with citations. For example, you can use www.easybib.com, a website that will automatically generate citations for you.

SAMPLE BIBLIOGRAPHICAL REFERENCES

Kaufman, L., and Edward Lodgen. "The Discovery of the Dietary Habits of New World Monkeys." *National Geographic* Aug. 2018: 128-133.

"Monkey." *LiveScience.* TechMedia Network, n.d. Web. 15 July 2018. <http://www.livescience.com/topics/monkey/>.

Rowe, N. *A Pictorial Guide to the Living Primates*. New York: Pogonias Press. 2012.

"Social Dominance in Apes". *All about primates*. 24 April 2016. <http://www.allaboutmonkeys.com/dominance/apes.html>.

Wise, C. "Bonobos". *The Complete Encyclopedia of Primates*. 3rd ed. 2019.

Model Excerpts For Paper 2

> Note: This was written by the instructor to provide an example of elements of a very good student paper.

Last Thursday, I went to the Los Angeles Zoo for the purposes of observing two primate species. I arrived at around eleven in the morning and decided to start my observations immediately since the day was rather warm and the sun was already getting hot. My first stop was at the chimpanzee exhibit. The enclosure, built in 1998, was quite impressive and probably covered several acres. There were about fifteen chimpanzees in the exhibit, a number that seemed appropriate since Wise has noted that chimps often sit together in large groups. A waterfall was placed in the middle of the exhibit. The chimpanzees appeared to like this feature since many sat near the cascading water, probably to keep cool. This was confirmed by a sign at the exhibit which noted that chimpanzees sometimes will stay near streams in the wild in order to keep their body temperatures down.

Notice that these two pieces of information are not facts that one would typically know. Since they came from an exhibit and one of the authors, references are needed. Simply stating the author's name (Wise) is acceptable.

I decided to observe a large male chimp who I named Bill for the purposes of this description. I chose this individual since he was sitting near the glass display and I could get rather close to him initially. My superior vantage point did not last long (only about 4 minutes, or 8% of the observation period) because soon after I arrived the zookeepers put some fruit out, a resource that is part of a typical chimpanzee diet (Galishoff). Other foods eaten by chimpanzees include leaves, herbs, honey, and some animal prey (Kaufman and Lodgen), but I did not witness them eating any of these items.

You should incorporate your research with your earlier observations. Do not simply staple the first two parts of your paper together. Notice this alternative style of reference that uses the parentheses with the author(s).

The zookeepers' actions of providing fruit resulted in some aggressive encounters between Bill and several other chimps. Although relatively rare, aggression is observed in chimpanzee groups (Social Dominance in Apes), which conforms to my observation that the aggression only lasted as long as the fruit, about four minutes in duration. "Social interactions between chimpanzee males can be described by the terms *dominance* and *submission*" (Rowe 185), and Bill certainly appeared to be the dominant individual in most of the bouts of aggression I observed. Of the time spent observing the chimps, aggression accounted for 12% of their time in total. After that, Bill spent some time grooming a smaller male chimp. Perhaps this was a form of reassurance that follows aggressive encounters, something that Rowe describes as typical. However, Bill had never been aggressive to the smaller male chimp that he was grooming, so perhaps there was another reason for this action.

In these source citations, other forms were used. One contains a reference to a website which has no author named. In the others, quotes were used for some information. This was done because the information was copied directly from the source. There is nothing wrong with this practice. This is not plagiarism since the source was cited. Notice that the page number is included when a direct quote is used.

The grooming that followed the aggression and a second bout later in the observation resulted in 14% of Bill's time being spent grooming....

PAPER ASSIGNMENTS: PAPER ASSIGNMENT 2

CHECKLIST

____ I read the information relating to this part of the paper included above.
____ I reviewed the comments for the first part of the paper.
____ I made revisions to the first part of my paper as suggested.
____ I incorporated my research into my present paper.
____ I included references to my sources within the body of my paper.
____ I used quotation marks when using information directly from sources.
____ I included all sources cited in a bibliography.
____ I included my thoughts about doing this part of the project into my paper.
____ I performed a spell-check on my paper.
____ I made both a hard copy and an electronic copy of my paper.
____ I submitted an electronic copy of my paper to the instructor.

STANDARDS

There are two major elements of this part of the paper:

1. The description of the behavior of two primates including time allocations from the first part of your paper; and
2. The integration of relevant facts discovered through your review of library and Internet sources.

There are also several minor elements:

1. The minor elements from the first part of the paper including a brief introduction, a species comparison, and a statement about your thoughts about the project;
2. References in your paper to your sources of information;
3. A bibliography including the sources you cited in the paper; and
4. More comments about what you thought about the fact-finding part of the project.

Below is a rough guide for the level of work expected to obtain various grades. Generally all of the elements in a particular group must be present to obtain the given grade. For example, a student who does a good job with all components, but completely separates the zoo observations from the outside facts would fall in the 60-point range since he or she would have "poor integration of the relevant facts."

120-100 Points	90 Points
• A thorough description complete with allocations	• Fairly thorough description with allocations
• A systematic integration of relevant facts discovered	• Moderate level of integration of facts discovered
• Complete incorporation of all minor elements	• Very good integration of all minor elements
• Few, if any, grammatical errors	• Few, if any, grammatical errors
80-70 Points	**60 Points or lower**
• Incomplete or poor time allocations	• Time allocations missing and/or poor description
• Incomplete integration of relevant facts discovered	• Poor integration of relevant facts discovered
• One or more missing minor elements	• A few minor elements missing or incomplete
• Average grammar	• Poor grammar or other omissions

Note: This sample, written by a student, is intended to provide an example of an excellent paper. Notice how well the zoo facts are integrated with the outside facts along with the time allocations. The paper may have gone a little far with the level of detail of outside information, but overall the effort is outstanding.

I spent a beautiful Monday afternoon at the Santa Ana Zoo at Prentice Park. The twenty-one acre zoo houses a collection of 17 different species of primates, including such species as the black-handed spider monkey, the golden lion tamarin, and the pygmy marmoset, among others. Five different endangered species also dwell in the Santa Ana Zoo. For my course project, I carefully evaluated my choices, and settled upon two endangered species: an ape, the white-handed gibbon, and a prosimian, the ring-tailed lemur.

White-handed gibbon, *hylobates lar* - The zookeeper at the Santa Ana Zoo was quick to let me know that their white-handed gibbon, Princess, is the smartest animal in their zoo, so I was eager to observe this intelligent species. The white-handed gibbon, like the gorilla, chimpanzee and orangutan, is an ape, not a monkey. In fact, this particular gibbon is the smallest primate in the ape family. Adult males weigh 10 to 20 pounds; females are slightly smaller in size. The chief characteristics that distinguish apes from monkeys are the absence of a tail, the more upright posture, and the high development of the brain (Boyd and Silk).

Hence the name of the primate, the palms of the hand and the soles of the feet are white in color. Their face is bare, as are the palms of their hands and the soles of their feet. Their fur is extremely dense, providing protection from rain. One square centimeter of skin has over 2,000 individual hairs (Swindler). Princess spent some 4% of her time, approximately two minutes, grooming her dense fur coat and licking her fur. She is black, but the white-handed gibbons come in two color forms, blond and black.

Interestingly enough, the white-handed gibbon is the most active of all gibbons. They move faster, more quietly, and farther each day than any other forest apes or monkeys. The animal's long arms and hook-like fingers contribute to its ability to swing through trees, a method of transportation known as brachiation. By swinging from branch to branch, these gibbons are able to change direction in flight and to catch a handhold in case they fall. The white-handed gibbon can easily leap a gap of 30 feet between one tree and another, but because they cannot swim, they avoid crossing open water. Adaptations include precision of movement, incredible eye-hand coordination, and dexterity (MacDonald). Princess's approximately 10 foot by 10 foot exhibition housed a complex network of branches, ropes, tree stubs and floral on which she brachiated and played. In fact, Princess spent a good 44% of the time brachiating and playing in the ropes and branches in her exhibition. An explanation for this prevalent behavior lies in the fact that brachiation is her method of transportation; in order to move, she must brachiate. The white-handed gibbon is an arboreal primate; it can be found in the upper canopy of the tropical forests of Southeast Asia, including Thailand, Malaysia, Indonesia, and Burma. In fact, the word "hylobates" in the name means "dweller in the trees."

One of the most interesting features of this primate is that the white-handed gibbon begins each morning with a whopping and piercing morning "song" which marks its territory. The white-handed gibbon is vigorously territorial, spending up to one half hour each morning calling, displaying, and marking its territory. Flannery mentions that the function of calling is territorial, but the call also serves to reinforce the pair bond. Although the female begins the "duet", the male responds, exchanging "songs" and responding to each other's calls. Although I did not attend the zoo early enough to hear Princess's call, the zookeeper assured me that Princess began each fresh morning with a loud and booming song.

The white-handed gibbon is a monogamous species that forms small groups consisting of one mated pair and their offspring; sometimes they live in family troops of 10 or more members within a 30-100 acre territory (Nowak). There is almost no sexual dimorphism, and males are not socially or physically dominant over females. The white-handed gibbon has no fixed season for breeding; the gestation period lasts around seven months, and infants are weaned within a two-year period. The young clings to the mother night and day, and at six months, it begins to brachiate on its own. When the animal reaches sexual maturity in 6 to 10 years, it meets other gibbons in common feeding groups, where after courtship, new family groups are formed. Princess, the white-handed gibbon at the Santa Ana Zoo, had a mate, but unfortunately, the male gibbon passed away only a month ago, leaving Princess depressed,

morose and inactive. The death certainly took its toll on Princess, for she spent well over 21 minutes merely resting and sleeping, spending a staggering 42% of her time idol. She did not engage any time playing with her mate or being aggressive, for she was the only creature in her exhibition.

Living in the upper canopy of tropical rainforests, the white-handed gibbon feeds predominantly on ripe fruits, although they also eat leaves, young plant shoots, flowers, small insects, and birds. Their zoo diet consists of fruit and plants; Princess, in the 10% of the time that she spent eating, ate figs (a favorite food) and various plants and leaves.

The white-handed gibbon is considered an endangered species, and is drastically declining in numbers for various reasons. As man enters its forested territory, he often "kills the mothers in capturing the young for a lucrative pet market" (MacDonald). In many Asian countries, it is "fashionable" to own a primate; this fad has led to the death of many gibbons (Nowak). In some areas, these primates are hunted for meat. However, the greatest threat to the white-handed gibbon is deforestation. Rainforests are disappearing at an alarming rate due to agricultural development, leaving these species with an even small region in which to live. White-handed gibbons retain only 10% of their original habitat in protected reserves (MacDonald).

Ring-tailed lemur, lemur catta - The Santa Ana Zoo houses a complete exhibition of ring-tailed lemurs, which includes seven of these animals. I was first drawn to these animals by their beautifully-colored rings on their tails, hence the name, 'ring-tailed lemur'. The word 'lemur' actually means 'ghost' in Madagascar, a word fitting for those mysteriously-looking creatures. I choose a specific lemur, whom I will call Spot, that spent a lot of his time away from the other lemurs; I observed this particular creature.

Unlike the white-handed gibbon, Spot and his family and friends lived in a huge exhibition that was equipped with numerous branches, logs, trees, and even a full waterfall with rocks. Later, I learned that true lemurs do not swim well and seldom enter water, so the presence of a waterfall was a bit uncanny. Many of the lemurs hung out in pairs or small groups, so it was odd for Spot to be alone. In fact, the ring-tailed lemurs are very sociable, living in large groups of up to 25 animals, with females dominant to males. Female dominance is unique to prosimians (Boyd and Silk). In these groups, ring-tailed lemurs have distinct hierarchies that are enforced by frequent and aggressive confrontations among members. During the times when Spot hung out with his family and friends, he spent 11 minutes or 22% of the time engaging in physical aggression with other lemurs. For 14 minutes or 28% of his time, Spot played by himself. Among those 13 minutes, six minutes were spent playing with his ringed tail, and seven minutes were spent climbing the gates of the exhibition.

Most lemurs are the size of a house cat, weighing about 6 to 8 pounds. Many ring-tailed lemurs are white-faced, black-eyed, and short-necked. They also have black, pointed muzzles, which is typical among the various species of lemur. The majority of their fur is a soft, light gray color, with variations in hue. The most distinguishing characteristic of this species is their ringed-tails, which are striped with 13 alternating blank and white bands, giving this lemur its common name.

While observing the ring-tailed lemurs at the Santa Ana Zoo, I noticed that Spot groomed himself in a strikingly unique way. Spending nearly 8% of the time grooming, Spot, as well as all prosimians, have six lower teeth that stick straight out from their jaw. These teeth form a sort of tooth comb that the animals use to groom their fur, as well as the fur of other members of their social group.

Ring-tailed lemurs are found solely on the island of Madagascar, an island off of southeast Africa. They live in arid and open areas and forests; they spend about 40% of their time on the ground. They typically walk on the ground or climb on large limbs in the trees. This preference differentiates them from other lemur species, which prefer forested areas and trees. I observed this phenomenon in the ring-railed lemur exhibition at the Santa Ana Zoo; the lemurs spent nearly equal amounts of time in the branches and leaves of the trees and on the ground. Spot, in particular, spent about 28% of his time merely walking, resting, and relaxing, both on the ground as well as in trees. I further divided this time into time on the ground and time in the trees: Spot spent approximately 12% of the time in trees and branches, and 16% of the time on the ground, on small logs, and on the rocks near the waterfall.

Ring-tailed females usually first give birth at three years of age, and produce offspring annually thereafter. The mating season is extremely seasonal, beginning in mid-April in the wild. The gestation period lasts about four months. Single infants are the common, but twins are a frequent sight in ringtail troops when food is plentiful. Initially, infants cling to their mothers' bellies, but after three weeks, they will take their first steps away from their mothers. Over the next five months, infants will spend increasing amounts of time on their own—returning to mom only to nurse or sleep—until they are finally weaned after six months time. Interestingly enough, females remain in

the same group for their entire lives, while males commonly change groups upon reaching sexual maturity. An average lifespan lasts for 20 to 25 years.

Ring-tails are the only true lemurs with 'stink' glands; they use their glandular secretions to mark their territory. The wet nose is associated with a keen sense of smell. Females have a large gland on the inside of each wrist, and males have one on each armpit. Male-to-male confrontations often involve stink-fights; they draw their tails through both armpits, smearing the scent throughout the fur (Anderson). Their tails are also used as flags, both for communication and protection. While observing these creatures, I never suspected any foul smell or scent, probably because there was no reason for competition or aggression, since all of the ring-tailed lemurs belonged to the same community.

Lemurs are primarily diurnal, avoiding activity only during the hottest parts of the day. They seek sunlight, especially in the morning, and sunbathe. However, although diurnal, lemurs have specialized night vision involving a reflective tapetum in the eyes' retina. The eyes "glow" in dim light when the viewer sees the reflected light pass back out of the eye (Allaby). The ring-tailed lemur's diet consists mainly of fruits, leaves, flowers, plants, and small insects. In the middle of my 50-minute observation period, a zookeeper entered to feed these creatures. Upon placing some branches and leaves in the exhibition, the zookeeper left, and all of the ring-tailed lemurs flocked to the treasure; it was quite a sight to see. Spot spent approximately seven minutes, or 14% of his time eating foliage and leaves. Many times, he did this sitting in the cracks of the trees.

Ring-tailed lemurs are in danger in Madagascar mainly because of deforestation for the timber industry. Their lands are rapidly being "converted to farmland, overgrazed by livestock, and harvested for charcoal production" (Burnie). As a result, at least 14 lemur species have become extinct. Frequently enough, ring-tailed lemurs are hunted for food or kept as pets. Although ring-tails are found in several protected areas in southern Madagascar, the level of protection varies widely in areas. Ring-tailed lemurs breed very well in captivity; over 1000 can be found in various zoos around the world.

I did experience some slight difficulties while conducting this project, especially during the time allocations. What if the gibbon spent 15 seconds grooming, and then proceeded to spend 5 seconds brachiating, and finished his minute by eating a leaf? How was I to divide this time? I solved this problem by splitting each of my 50 minutes into 50 individual time slots; in each time slot (which was divided into seconds), I noted the percentage of time spent in that activity (15% grooming, 5% brachiating, 40% eating). Although this procedure entailed more writing, focus, and calculations, I found my recordings to be fairly accurate.

One of the most eye-opening and fascinating facts that I learned through conducting this study is the profound variation among various animals. Although the white-handed gibbon and the ring-tailed lemur are both primates, the two species have many differences—physically, socially, geographically and intellectually. Although both animals are rather small in size, the gibbon is a lesser ape, and thus possesses no tail. However, the ring-tailed lemur's distinguishing feature is its beautifully-ringed tail. Furthermore, white-handed gibbons live in monogamous groups, whereas ring-tailed lemurs live in multi-male, multi-female groups. Despite the fact that both species live in tropical forests, the white-handed gibbon lives in Southeast Asia, while the ring-tailed lemur can be found only in Madagascar. In both these areas, both species are considered endangered; this study undoubtedly has given me a greater appreciation of each animal in its entirety for its unique characteristics and features.

Works Cited

Allaby, Michael, ed. *The Concise Oxford Dictionary of Zoology*. Oxford University Press: London. 1992.

Anderson, Rebecca. *Lemur catta: narrative*. 25 June 2003. <http://animaldiversity.ummz.umich.edu/accounts/lemur/ l._catta$narrative.html>.

Boyd, Robert and Joan B. Silk. *How Humans Evolved*, 3rd ed. W.W. Norton & Company, Inc.: New York. 2003.

Burnie, David. *Primate*. Microsoft® Encarta® Online Encyclopedia 2003. 3 July 2003.
        <http://encarta.msn.com>.

Flannery, Sean. *White-handed Gibbon* (Hylobates lar). 3 July 2003.
        <http://www.primate.wisc.edu/pin/factsheets/hylobates_lar.html>.

MacDonald, David W. The Encyclopedia of Mammals. Checkmark Books: Oxford. 1995.

Nowak, Ronald M. Walker's Mammals of the World, 6th ed, Vol I. Johns Hopkins University Press: Baltimore. 1999.

Swindler, Ray Daris. *Introduction to the Primates*. University of Washington Press: Washington. 1998.

Made in the USA
San Bernardino, CA
05 January 2020